Robert Stewart

Handbook of the Torquay Flora

Comprising the flowering Plants and Ferns growing in and around Torquay

Robert Stewart

Handbook of the Torquay Flora
Comprising the flowering Plants and Ferns growing in and around Torquay

ISBN/EAN: 9783337107345

Printed in Europe, USA, Canada, Australia, Japan

Cover: Foto ©ninafisch / pixelio.de

More available books at **www.hansebooks.com**

HANDBOOK

OF

THE TORQUAY FLORA;

COMPRISING THE

FLOWERING PLANTS AND FERNS GROWING
IN AND AROUND TORQUAY,

WITH THEIR RESPECTIVE HABITATS.

BY

ROBERT STEWART, M.R.C.S.

TORQUAY:
E. CROYDON, ROYAL LIBRARY.
LONDON:
HAMILTON, ADAMS, AND CO., PATERNOSTER ROW.
1860.

TO

PERCIVALL HARE EARLE, ESQ.,

IN REMEMBRANCE OF HIS COMPANIONSHIP DURING MANY

PLEASANT BOTANICAL RAMBLES,

This Volume is Dedicated,

WITH EVERY FEELING OF AFFECTION AND ESTEEM,

BY HIS SINCERE FRIEND

THE AUTHOR.

PREFACE.

The following work merely professes to be a List of the various habitats of the Flowering Plants and Ferns of Torquay and its neighbourhood, and lays no claim whatever to be considered in a scientific point of view; it is only designed to facilitate the researches of those who are desirous of botanizing in our beautiful vicinity, by pointing out, as clearly as possible, the particular localities in which the majority of our wild plants are to be found.

I have followed the classification and arrangement of Hooker and Arnott's 'British Flora' (seventh edition), and have relied, for the habitats of the plants enumerated, chiefly on my own knowledge of localities after nearly twenty years' botanizing in the neighbourhood; but I have availed myself also of a large collection of Devonian plants presented to the Torquay Natural History Society by C. E. Parker, Esq., and of a small but valuable collection of the rarer Flowering Plants, and a beautiful collection of Ferns, presented to the same Institution by Miss A. Griffiths.

Although my more immediate object has been to give the several habitats in and near Torquay, still I have deemed it advisable to take a rather wider range, and to specify some more distant stations of plants, but in those places almost exclusively to which easy excursions can be made; I have therefore included all the localities with which I am acquainted, within a circuit of twenty-five miles of Torquay.

After the name of each plant, I have stated the nature of the situations in which it is most likely to be found, followed, in most cases, by a short description of its more obvious characters; and then I have given a list of its several local habitats, the number of the plate in Sowerby's 'English Botany' which represents it, and the time of its flowering, but I have not thought it necessary to enter into any minute botanical details, inasmuch as every botanist will possess one, at least, of the many valuable English Floras which have of late years been published,* each of which contains all the requisite help for ascertaining species.

When first botanizing in this locality, I felt greatly the want of a similar work to the present, and I now confidently trust that my humble labour may prove of service to those who are anxious to make themselves acquainted with the wild plants which grow in the neighbourhood of Torquay.

In conclusion, I have only to state that I shall feel grateful to those botanists who will communicate to me any well authenticated habitats, within the prescribed area, which I have not specified.

R. STEWART.

3, *Park Place, Torquay.*

---

* The best modern Manuals are the following :—Hooker and Arnott's 'British Flora;' Babington's Manual; and Bentham's 'Handbook of the British Flora.'

ABBREVIATIONS USED.

E. B. ....English Botany.
E. B. S. ....Supplement to English Botany.
Bab................Babington's Manual.
Benth..............Bentham's Handbook.
Fl. D. ....Flora Devoniensis.
Linn. ....Linnæus.
Sm. ....Smith.

A..................Annual.
B..................Biennial.
P..................Perennial.
Sh. ....Shrub.
T..................Tree.

I. II. III. etc. represent the months of flowering, as January, February, March, etc.

HANDBOOK

OF THE

TORQUAY FLORA.

CLASS I. DICOTYLEDONOUS OR EXOGENOUS PLANTS.

SUB-CLASS I. THALAMIFLORÆ (ORD. I.–XXIII.)

ORD. I. RANUNCULACEÆ.

CLEMATIS. TRAVELLER'S JOY.

C. Vitalba (*common T.*)—A climbing plant with a woody stem, abundant in woods and hedges. Leaves pinnate, leafstalks twining; segments usually 5, ovate and stalked. The young branches twist and twine over hedges and trees, sometimes completely covering them up. Flowers greenish-white, succeeded by an abundant silky and feathery seed, perfectly white, at a distance having the appearance of snow. Very plentiful on hedges and in woods about Torquay and Marychurch. (E. B. t. 612.) Sh. VI.

THALICTRUM. MEADOW-RUE.

T. minus (*lesser M.*)—Stony pastures, and in dry situations in limestone countries. A pretty, graceful-looking plant, with a zigzag stem, about 1 foot high. Leafstalks three or four times divided, with numerous small roundish or broadly wedge-shaped leaves. Flowers a pale greenish-yellow, the calyx tinged with pink. Babbicombe Down, on some projecting rocks near a steep descent, just beyond the first gate. (E. B. t. 11.) P. VI. VII.

ANEMONE. ANEMONE.

A. nemorosa (*wood A.*)—Moist woods and pastures and on high mountains. Plant consisting of 2 or 3 root-leaves and a single flower-stalk. Leafstalks long, with 3 ovate leaflets, lobed, or sometimes divided into 3 similarly shaped segments. Flower-stalk from 3 to 8 inches high; flowers large and white, tinged with purple on the outside. Bradley Woods. Near Totness. Berry Pomeroy. Lane leading to Gidleigh, near Chagford. (E. B. t. 355.) P. III.-V.

RANUNCULUS. CROWFOOT, SPEARWORT.

1. **R. circinatus** (*rigid-leaved Water C.*)—Lakes, ponds, and ditches. Closely resembling the following, and doubtful whether it should be considered as distinct from it. All the leaves with shorter segments, spreading in all directions. Flowers large. Near Barton. (E. B. S. t. 2869.) P. VI.-VIII.

2. **R. aquatilis** (*common Water C.*)—Ponds, lakes, and ditches. Plant either creeping in mud or floating in water, very variable in appearance. When floating, the lower and sometimes all the leaves remain under-water; leaves divided into 3 or 5 wedge-shaped or rounded lobes, those submerged linear; flowers white, with their petals larger than the calyx. Forde bog, near Newton. Goodrington marsh. River Dart. (Bovey Heath, near the coal-pit, *Fl. D.*) (E. B. t. 101.) P. VI.-VIII.

3. **R. hederaceus** (*Ivy C.*)—In wet places and shallow pools of water. All the leaves spread on the mud or floating, rounded or kidney-shaped, smooth and succulent, on long leafstalks. Flowers small and white. Cockington. Paignton. Forde bog, near Newton. (E. B. t. 2003.) P. VI.-IX.

4. **R. Flammula** (*lesser Spearwort.*)—Sides of lakes and in ditches. Stem more or less prostrate at the base, rooting at the joints and sending up stems about a foot high, with long lance-shaped leaves; flowers on long flower-stalks, of a bright yellow hue. Forde bog, near Newton. Bovey Heath. (E. B. t. 387.) P. VI.-VIII.

5. **R. Ficaria** (*Pilewort C., lesser Celandine.*)—On banks, in pastures, woods and bushy places, abundant. Plant with radical heart-shaped leaves, crenated, smooth, and shining. Flower-stem bearing one or two stalked leaves, flowers yellow, with an almost metallic lustre. Very common in the lanes about Torquay and Cockington. (E. B. t. 584.) P. III.-V.

6. **R. auricomus** (*wood C.*, *Goldilocks.*)—In woods, thickets, and bushy places. Plant erect and branched, usually about 8 inches or a foot in height. Root-leaves kidney-shaped, stalked; upper ones divided into 3, 5, or 7 linear toothed segments; calyx hairy, shorter than the petals; flowers large and yellow. Plant not acrid. Berry Pomeroy. (E. B. t. 624.) P. IV. V.

7. **R. sceleratus** (*Celery-leaved C.*)—On the sides of ditches and pools. A much branched and upright plant, sometimes growing to nearly 2 feet high, but generally under a foot. The lower leaves stalked, and divided into 3 bluntly-toothed lobes; upper leaves sessile, consisting of 3 linear segments. Flowers very small and numerous, pale yellow. Side of the ditch near Torre Abbey. Paignton. Dawlish. (E. B. t. 681.) A. VI.-IX.

8. **R. acris** (*upright meadow C.*)—In meadows and pastures. This plant varies in height from 6 inches to 2 or 3 feet. Leaves 3 times divided, with deeply and acutely cut segments; upper leaves linear. Flowers large and bright yellow, on long panicled stalks. Calyx spreading, but shorter than the petals. Common in pastures everywhere. (E. B. t. 652.) P. VI. VII.

9. **R. repens** (*creeping C.*)—In pastures, very common. Roots creeping. Stems about 1 foot high; leaves with 3 stalked leaflets, usually 3-lobed and cut. Flowers large and yellow. Seed-vessel hairy. Very abundant in pastures. (E. B. t. 516.) P. V.-VIII.

10. **R. bulbosus** (*bulbous C.*)—In pastures, meadows, and waste places. Stem from 6 inches to a foot high, thickened at the base into a kind of bulb, hairy. Leaves smaller than those of *R. repens*, divided into 3 segments which are more or less cut; the upper leaves cut into linear segments. Stem many-flowered; flowers yellow. Torquay. Marychurch, etc., frequent. (E. B. t. 515.) P. V.

11. **R. hirsutus** (*pale hairy C.*)—In fields and waste places. Plant much branched, 6 inches to 1 foot high; much like the last, but with leaves and flowers smaller, and the latter of a paler yellow. Torre Abbey ditches. Marychurch. (E. B. t. 1504.) A. VI.-X.

12. **R. parviflorus** (*small-flowered C.*)—In cornfields and waste places. Plant from 2 or 3 inches to a foot high, with a weak, decumbent or ascending stem; whole plant hairy. Leaves nearly round, on long stalks, the lower ones 5-lobed and cut, the upper 3-lobed. Flowers yellow, small; petals not longer than the calyx. Road to Meadfoot. Pastures, etc., above Hope's Nose. Warberry Hill, etc. (E. B. t. 120.) A. V. VI.

13. **R. arvensis** (*corn C.*)—Cornfields. Stem erect, branching, pale green, 6 to 18 inches high. Leaves divided into long

narrow segments. Flowers pale yellow and small. Readily distinguished by its spinous fruit. Cornfields in the neighbourhood of Exeter, Withecombe Wood, near Exmouth, *Fl. D.* (E. B. t. 135.) A. v.

CALTHA. MARSH MARIGOLD.

C. palustris (*common M.*)—Marshy places, common. Stem rooting, about a foot long; leaves nearly all radical, on long stalks, kidney-shaped or roundish. Flowers very large and handsome, of a bright golden yellow. Behind Torre Abbey. Paignton. Kingskerswell. Forde bog. Totness. (E. B. t. 2175.) P. III.-VI.

HELLEBORUS. HELLEBORE.

1. **H. viridis** (*green H.*)—In woods, thickets, and hedges, growing frequently about old walls and ruined houses. Plant about 1 foot or 18 inches high. The root-leaves large, on long, broad stalks, divided into from 7 to 11 narrow lanceolate and serrated segments, 3 or 4 inches long; upper leaves sessile. Flowers 3 or 4, drooping, of a pale sickly green. Chelstone, near Torquay, in an old orchard. (E. B. t. 200.) P. III. IV.

2. **H. fœtidus** (*stinking H.*)—In stony pastures and thickets. Flower-stem more than a foot high. Leaves not all radical, but growing from the base of the stem in a large and thick tuft, having narrower segments. Flowers more numerous than in *H. viridis*, growing more in clusters, pale green, sometimes with a purple tinge. Torquay. Miss A. Griffiths. (E. B. t. 613.) P. II.-IV.

AQUILEGIA. COLUMBINE.

A. vulgaris (*common C.*)—Woods and coppices. Low leaves growing in a large cluster, on long stalks, two or three times divided; segments broad and having 3 lobes. Flowering stems bearing a loose panicle of large, drooping, blue, dull purple, or white flowers. Near Hope's Nose. Kerswell Down. Woods at Ipplepen. Chudleigh. Gidleigh, near Chagford. (E. B. t. 297.) P. V.-VII.

ORD. II. **BERBERIDACEÆ.**

BERBERIS. BARBERY.

B. vulgaris (*common B.*)—Coppices, woods, and hedges. A small shrub, growing to about 6 or 8 feet high, branches armed with sharp thorns at the base of the leaf-tufts. Leaves ovate and sharply toothed. Flowers yellow, hanging in graceful clusters. Berries red, very acid. Shiphay, near Torquay. Copse by the brook at Chudleigh. (E. B. t. 49.) Sh. v. vi.

ORD. III. **NYMPHÆACEÆ.**

NYMPHÆA. WHITE WATER-LILY.

N. alba (*great W.*)—Lakes and pools. A beautiful aquatic plant, with large heart-shaped leaves, and pure white flowers 3 or 4 inches in diameter, lying on the surface of the water. Given in Fl. D. as growing in marshes and canals at Powderham, but most probably not wild. (E. B. t. 160.) P. vii.

NUPHAR. YELLOW WATER-LILY.

N. lutea (*common Yellow W.*)—In the same situations as *Nymphæa alba*, but more common. Leaves heart-shaped. Flowers yellow and rising on stalks some inches out of the water; calyx large, but the petals of the flower small and numerous. Fruit roundish. In the Clyst river, near Bishop's Clyst Bridge, *Fl. D.* (E. B. t. 159.) P. vii.

ORD. IV. **PAPAVERACEÆ.**

PAPAVER. POPPY.

1. **P. hybridum** (*round rough-headed P.*)—Waste and cultivated places, rather rare. Leaves stalked, once or twice pinnate, with stiff and short segments. Flowers rather small, of a purplish red. Seed-capsules, round, covered with stiff spreading bristles. Fields about Torquay. In a field near Dawlish, *Fl. D.* (E. B. t. 43.) A. v.–vii.

2. **P. Argemone** (*long prickly-headed P.*)—Corn-fields and

waste places. A weak and small plant, with few and narrow leaves. Flowers small with narrow petals, scarlet-red rather than crimson, often having a dark spot upon them. Capsule hairy with erect bristles. Fields at North Bovey. Mount Pleasant, above Exmouth Warren. Fields by the Exe, near Exeter, *Fl. D.* (E. B. t. 643.) A. v.–vii.

3. **P. dubium** (*long smooth-headed P.*)—Waste places and cornfields, rather common. Stem 1 to 2 feet high. Leaves singly or doubly pinnatifid, sessile, with very narrow lobes. Flowers large, with broad petals, light scarlet. Fruit oblong and smooth. Cornfields at Marychurch. Newton. (E. B. t. 644.) A. v.–vii.

4. **P. Rhœas** (*common red P.*)—In waste and cultivated places, cornfields. Plant erect and branched, 1 or 2 feet high, whole plant rough with spreading hairs. Lower leaves large and stalked, once or twice divided, with the lobes lance-shaped and pointed; upper leaves sessile, deeply cut, with serrated segments. Flowers large, with broad overlapping petals, deep crimson. Capsule round and smooth. Cornfields about Torquay, etc. (E. B. t. 645.) A. vi. vii.

GLAUCIUM. HORNED POPPY.

G. luteum (*yellow Horned Poppy.*)—On sandy seashores. A stout plant, with hard, spreading, smooth branches. Leaves thick and fleshy; lower ones on stalks and divided, bearing short thick hairs; upper leaves embracing the stem, wavy and smooth. Flowers on short stalks, large and bright yellow, succeeded by curved pods from 6 inches to a foot in length, which ar surmountede by the lobes of the stigma. Meadfoot Cliffs, Torquay. Paignton. (E. B. t. 8.) B. vi.–viii.

CHELIDONIUM. CELANDINE.

C. majus (*common C.*)—On roadsides, waste places, and on old walls. Plant from 1 to 2 feet high; stems slender, branching, giving out a yellow fetid juice when broken. Leaves thin and delicate. Leaves pinnate; segments ovate, coarsely lobed. Flowers small and yellow, 5 or 6 together in an imperfect umbel; seed-vessel sticking up from the centre of the flower. Pod long, swollen in the centre and tapering at each end. Chelston. Shiphay, on old walls. (E. B. t. 1581.) P. v.–viii.

Ord. V. FUMARIACEÆ.

FUMARIA. FUMITORY.

1. **F. officinalis** (*common F.*)—In dry fields and by roadsides. A delicate plant, of a pale green colour, varying much both in size and appearance. Leaves smooth and very much divided, generally 3-lobed; lobes varying in shape from linear to oblong. Flowers red, small. Nuts very blunt. Meadfoot cliffs, etc. (E. B. t. 589.) A. v.-ix.

2. **F. capreolata** (*rampant F.*)—In cornfields, gardens, hedges, and roadsides. A large luxuriant form, very variable, much bolder in all its parts than the last, often spreading out to the length of 2 feet or more. Flowers whitish or pale red. Nuts nearly globular. Babbicombe. Teignmouth. (E. B. t. 943.) A. vi.-ix.

CORYDALIS. CORYDALIS.

1. **C. lutea** (*yellow C.*)—In stony places and on old walls. Stem angular, erect, 6 or 8 inches high. Leaves delicate, almost transparent, pale green, divided into a great number of ovate or wedge-shaped segments. Flowers in short racemes, pale yellow. Pod about a fourth of an inch long. Cockington, on old walls. Shiphay. (*Fumaria*, E. B. t. 588.) P. v.-viii.

2. **C. claviculata** (*white climbing C.*)—In stony and bushy places. Plant with slender much-branched stems 1 or 2 feet long. Leafstalks ending at the terminal leaf in a delicate tendril, which enables the plant to climb among stones and bushes; leaves much divided into small oblong-toothed segments. Flowers small and cream-coloured. Meadfoot cliffs. Holne Chase, etc. *Fumaria*, Linn. (E. B. t. 103.) P. vi. vii.

Ord. VI. CRUCIFERÆ.

Subord. I. *PLEURORHIZÆ*.

Tribe I. Arabideæ.

CHEIRANTHUS. WALLFLOWER.

C. Cheiri (*common W.*)—On old walls and waste places. A stiff, hardy plant, with a woody-looking stem. Leaves numerous, narrow, pointed, and quite entire. Flowers large, of a deep orange-yellow, sometimes pale yellow, sweet-scented   Wall of the old manor-house, Torquay.   Wall at Chelston, etc.   *C. fructiculosus*, Linn. (E. B. t. 1934.)   P. v. vi.

BARBAREA.   WINTER-CRESS.

1. **B. vulgaris** (*bitter W., yellow Rocket.*)—Pastures and hedges, common. A stout, green, smooth, and slightly branched plant, from 1 to 2 feet high. Lower leaves lyrate, with blunt segments; upper ones mostly pinnate, with blunt terminal lobes. Flowers small and yellow, the lower ones falling off before the upper flowers open. Pods numerous, erect. *Erysimum Barbarea*, Linn. Frequent everywhere. (E. B. t. 443.)   B. or P. v.–viii.

2. **B. præcox** (*early W.*)—Pastures and hedges. Agreeing very closely with the last, but more slender in all its parts. Style shorter and thicker. Thought by Bentham to be merely a variety. Chelston. Teignmouth. (*Erysimum*, E. B. t. 1120.)   B. v.–vii.

ARABIS.   ROCK-CRESS.

A. hirsuta (*hairy R.*)—On walls, banks, and rocks. Plant from 10 to 15 inches high; stem erect, and rough with short hairs; lower leaves spreading, oblong and slightly toothed, upper leaves for the most part clasping the stem. Flowers small and white; pods long and erect. Meadford Cliffs. Park Hill. Chapel Hill. (E. B. t. 587.)   B. vi.–viii.

CARDAMINE.   BITTER-CRESS.

1. **C. pratensis** (*common B., Ladies' Smock, Cuckoo-flower.*) —In moist meadows. Stems nearly a foot high; leaves pinnate, radical ones with oval or nearly round segments; stem-leaves with oblong or linear segments. Flowers light pink, large and handsome; seed-pods long and upright. Meadows near Torre Abbey. Meadows at Paignton. Forde bog, Newton. (E. B. t. 776.) P. iv.–vi.

2. **C. hirsuta** (*hairy B.*)--In moist shady places, ommon. Plant from 6 inches to a foot high. Lower leaves pinnate, with ovate or rounded segments, angularly cut, upper ones more entire. Stem usually hairy, but sometimes nearly or quite smooth.

Flowers small and white. Pods in a loose cluster. Common in the neighbourhood of Torquay. (E. B. t. 492.) A. III.-VIII.

NASTURTIUM. WATER-CRESS, YELLOW-CRESS.

1. **N. officinale** (*common Water-Cress.*)—Brooks and rivulets, widely distributed. Stem much branched, creeping or floating in the water. Radical leaves large, of from 5 to 7 distant leaflets, the terminal one broad and rounded; stem-leaves with the leaflets smaller and much closer together, ovate or linear-oblong, but the end leaflet always largest and rounded. Flowers small and white. Seed-pods on spreading footstalks, curving slightly upwards. Much esteemed as a salad. Brooks and streams in the neighbourhood. (E. B. t. 855.) P. v.-x.

2. **N. sylvestre** (*creeping Yellow-Cress.*)—In wet places and on riverbanks. Plant with a creeping stem and ascending flowering branches. Leaves deeply pinnate; leaflets lanceolate and cut, the leaflets of the upper leaves however much narrower and entire. Flower-stem angularly waved; flowers small and yellow. Pod smaller and more slender than that of the common Water-Cress. Bovey stream, by Woodford Bridge, *Fl. D. Sisymbrium*, Linn. (E. B. t. 2324.) P. VI.-VIII.

3. **N. terrestre** (*marsh Yellow-Cress.*)—In muddy and watery places. Resembles the last, but not so tall and more slender; the leaves have broader lobes and rather more toothed; flowers and pods both smaller. Side of the river at Exwick, *Fl. D. Sisymbrium*, Linn. (E. B. t. 1747.) A. VI.-X.

TRIBE II. ALYSINEÆ.

ARMORACIA. HORSE-RADISH, WATER-RADISH.

1. **A. amphibia** (*great Water-Radish.*)—In moist meadows and sides of rivers. Plant rising 2 or 3 feet high. Leaves oblong, pointed and deeply serrate. Flowers yellow, with petals much longer than the calyx. Seed-pouch oval. River Exe, near the village of Wear, *Fl. D. Sisymbrium*, Linn. (E. B. t. 1840.) P. VII.-VIII.

2. **A. rusticana** (*common Horse-Radish.*)—Waste places and pastures. A well-known plant, but generally the outcast from gardens. Many places about Torquay. *Cochlearia armoracia*, Linn. (E. B. t. 2323.) P. v.

COCHLEARIA. SCURVY-GRASS.

1. **C. officinalis** (*common S.*)—On the seacoast, in a stony or sandy soil, frequent. A low, smooth, and somewhat fleshy plant, with stalked roundish or kidney-shaped lower leaves, and upper leaves sessile and inclining to oblong, both being angularly toothed. Flowers at the ends of the branches, white. Pods nearly round. Babbicombe. Walls and rocks about Torquay. (E. B. t. 551.) The variety γ of Hooker and Arnott, *C. Danica*, grows on the cliffs at Meadfoot. (E. B. t. 696.) B. v.–viii.

2. **C. Anglica** (*English S.*)—Sides of rivers, and marshy places. Differing from the last in its smaller size, its smaller stalked and heart-shaped root-leaves, larger flowers and seed-pods. Topsham marshes, *Fl. D.* (E. B. t. 552.) A. v.

DRABA. WHITLOW-GRASS.

D. verna (*common W.*)—On rocks, walls, and dry banks, very frequent. A very small plant, with a tuft of small oblong leaves, cut at their extremities, spreading closely on the ground, from the midst of which one or two naked flower-stalks arise, bearing a loose raceme of white flowers, succeeded by seed-pods on long stalks. Common everywhere about Torquay and Marychurch. (E. B. t. 586.) A. iii.–vi.

KONIGA. KONIGA.

K. maritima (*seaside K.*, or *sweet Alyssum.*)—In waste places and sandy pastures near the sea. Stems procumbent or ascending, nearly a foot long. Leaves narrow and lanceolate. Flowers white, smelling like honey. Pods of a swelling oval shape. Can only be considered as a truant from gardens. Exmouth sands. (E. B. t. 1729.) P. viii. ix.

Tribe III. Thlaspideæ.

THLASPI. PENNY-CRESS.

1. **T. arvense** (*field P.*, or *Mithridate Mustard.*)—In fields and by the sides of roads, not very common. A smooth upright plant, from 6 inches to a foot high. Leaves a pointed oval, inclining to lanceolate; lower leaves stalked; upper, clasping the stem with angular auricles. Flowers white and very minute; seed-

pods very large and swelling, deeply notched at the top. Warberry Hill, Torquay. Chudleigh. Teignmouth. (E. B. t. 1659.) A. V.–VII.

2. **T. alpestre** (*alpine P.*)—Mountainous pastures in limestone districts. Whole plant about 6 inches high. Radical leaves nearly oval, stalked; stem-leaves narrow and clasping, with short auricles. Flowers larger than in the last species, and seed-pods much smaller. Meadfoot Cliffs. (Ilsington, *Fl. D.*) (E. B. t. 81.) P. VI.–VIII.

TEESDALIA. TEESDALIA.

T. nudicaulis (*naked-stalked T.*)—On sandy and gravelly banks and stony places. Leaves all radical, pinnate, with the end lobe larger and broadly oval. Flower-stems 3 or 4, leafless, rising 2 or 3 inches high, crowned with a cluster of small white flowers. When in seed the pods are in short racemes, and are nearly round, with a slight notch at their tops. Bickleigh High Tor. Near Fingle Bridge. Base of Middledon Down, near Chagford. Banks at Bovey Heathfield. (E. B. t. 327.) A. VI. VII.

TRIBE IV. CAKILINEÆ.

CAKILE. SEA-ROCKET.

C. maritima (*purple S.*)—On sandy sea-shores, frequent. Plant with loose straggling branches; leaves fleshy and pinnately divided. Flowers purplish, something like those of the Stock. Paignton Sands. Goodrington Sands. (E. B. t. 231.) A. VI. VII.

SUBORD. *NOTORRHIZEÆ.*

TRIBE V. SISYMBRIEÆ.

SISYMBRIUM. HEDGE-MUSTARD.

1. **S. officinale** (*common H.*)—Waste places and roadsides, common. Plant erect, about 1 foot high, more or less downy, with stiff and spreading branches. Leaves very much divided, with lanceolate lobes, the end lobe much longer than the lateral ones. Flowers yellow and very small. Pods long and tapering, and closely pressed to the stalk. Very common about Torquay, etc. (E. B. t. 735.) A. VI. VII.

2. **S. Sophia** (*fine-leaved H., or Flix-weed.*)—Similar local'ties to the last. Stem from 1 to 2 feet high, with a short hoary down. Leaves divided 2 or 3 times into many linear segments. Flowers small and yellow. Pods long and slender, standing out from the stalk. Waste places at Teignmouth, *Fl. D.* (E. B. t. 963.) A. VI.–VIII.

3. **S. thalianum** (*common Thale-Cress.*)—On old walls and dry banks. A slender branching plant about 6 inches high. Leaves nearly all radical, and spreading, coarsely toothed. Stem-leaves few and sessile. Flowers small and white. Pods slender, on slight stalks. Common on banks and walls. *Arabis*, Linn. (E. B. t. 901.) A. IV. V. and IX. X.

ALLIARIA. GARLIC-MUSTARD.

A. officinalis (*common G., Jack-by-the hedge, or Sauce-alone.*)—By waysides and on banks of hedges, very common. From 1 to 3 feet high. Lower leaves roundish and crenated, on long stalks; those of the stem on short stalks, heart-shaped, pointed at the extremities and coarsely toothed. Flowers white. Seed-pods on short stalks, the pod making an angular bend at its junction with the stalk. Whole plant, when bruised, smells strongly of garlic. Abundant along the Babbicombe road and Ansti's Cove lane. *Erysimum Alliaria*, Linn. (E. B. t. 796.) B. V. VI.

ERYSIMUM. TREACLE-MUSTARD.

E. cheiranthoides (*worm-seed T.*)—In fields and waste places. A stout, erect plant from 10 inches to 2 feet high. Leaves broadly lance-shaped, tapering at the base, pale green. Flowers light yellow, small. Pods long and numerous. Fields at Marychurch. Newton. Paignton. (E. B. t. 942.) B. VI.–VIII.

Tribe VI. Camelineæ.

Tribe VII. Lepidineæ.

CAPSELLA. SHEPHERD'S-PURSE.

C. Bursa-pastoris (*common S.*)—Cornfields, roadsides, and waste places. Whole plant hairy; varies in height from 3 inches to 2 feet. Root-leaves pinnatifid, with the end lobe triangular, all spreading on the ground; stem-leaves clasping, with short auricles. Flowers small and white. Pods flattened, heart-shaped. Very common in the neighbourhood. *Thlaspi*, Linn. (E. B. t. 1485.) A. III.-X.

LEPIDIUM. PEPPERWORT.

1. **L. campestre** (*common Mithridate P.*)—In cornfields and dry, gravelly places. Stem solitary and erect, about a foot high, generally branched in the upper part. Root-leaves stalked and oblong; upper leaves inclined to lanceolate, slightly toothed, clasping, with pointed auricles. Flowers small and white. Pods very numerous, on spreading stalks. Plentiful in fields about the neighbourhood. (E. B. t. 1385.) *Thlaspi*, Linn. A. V.-VIII.

2. **L. Smithii** (*smooth field P.*)—In hilly pastures and waste places. Much like the last, but with shorter and more numerous stems. Leaves more hairy, and flowers larger. Seed-pod smooth. Meadfoot Cliffs. Warberry Hill. *Thlaspi hirtum*, Sm. (E. B. t. 1803.) P. VI.-VIII.

SENEBIERA. WART-CRESS.

1. **S. Coronopus** (*common W., Swine's-Cress.*)—In cultivated fields and waste places. Stems first forming a short tuft; but after flowering, spreading along the ground. Leaves 2 or 3 times pinnately divided, segments nearly linear. Flowers few, small, and white, in lateral axillary clusters. Seed-pouches large, in crowded clusters. Meadfoot Cliffs. Warberry Hill. Berry Head. Chudleigh. *Coronopus Ruellii*, Sm. (E. B. t. 1660.) A. V.-IX.

2. **S. didyma** (*lesser W.*)—Waste grounds near the sea. Somewhat like the last, but more slender. Leaves much more divided and more crowded. Flowers smaller. Frequent everywhere. *Coronopus*, Sm. (E. B. t. 248.) A. VII.-IX.

TRIBE VIII. ISATIDEÆ.

SUBORD. III. *ORTHOPLOCEÆ.*

TRIBE IX. BRASSICEÆ.

BRASSICA. CABBAGE, TURNIP, NAVEW.

1. **B. oleracea** (*sea Cabbage.*)—Cliffs near the sea. Leaves

smooth, waved and lobed; lower leaves lyrate, stalked; upper leaves oblong and sessile. Flowers large and yellow. Seed-pods long and erect. The origin of our garden Cabbage. Cliffs above Ansti's Cove. Dartmouth. (E. B. t. 637.) B. VI.–VIII.

2. **B. Napus** (*Rape or Cole-seed*) and **B. rapa** (*common Turnip*) are mere varieties dependent upon cultivation of **B. campestris** (*common Wild Navew*). A plant varying in height from 1 to 2 feet, with lower leaves lyrate and toothed, the end lobe large; somewhat hairy; upper leaves heart-shaped, tapering to a point and clasping the stem; auricles rounded. Flowers bright yellow. Pods like those of the Cabbage. Marychurch. Warberry Hill. Fields at Moreton, etc. (E. B. t. 2146, 2234, and 2176.) A. or B. V.–VII.

SINAPIS. MUSTARD.

1. **S. nigra** (*common M.*)—Under hedges and in waste grounds. Plant from 3 to 4 feet high. Lower leaves lyrate and rough, the end lobe large; upper leaves entire, lance-shaped. Flowers yellow. Pods pressed close to the stem. Our table Mustard is produced from the seeds. Common about Torquay, etc. (E. B. t. 969.) A. VI.–VIII.

2. **S. arvensis** (*wild M., Charlock.*)—Too common in cornfields. A coarse-looking plant, from 1 to 2 feet high, bristly. Lower leaves stalked, inclining to lyrate; upper leaves sessile. Flowers large and yellow. Pods smooth, swelling and knotty, beaked. Abundant everywhere. (E. B. t. 1748.) A. VI.–VIII.

3. **S. alba** (*white M.*)—Fields and waste places. Stem from 12 to 18 inches high, smooth, or with a few spreading hairs. Leaves lobed irregularly, pinnate or lyrate. Flowers large, yellow. Pods hairy. Very frequent. (E. B. t. 1677.) A. VII.

DIPLOTAXIS. ROCKET.

D. tenuifolia (*wall Rocket.*)—On walls and waste and rubbishy banks. From 1 to 2 feet high, loosely branched, quite smooth. Leaves a yellowish green, very variable in their divisions, usually lanceolate, with acute pinnate divisions; uppermost leaves often nearly entire. Flowers large and yellow. Pods slender and spreading, in a loose cluster, containing two distinct rows of seeds. Old walls about Exeter, *Fl. D. Sisymbrium*, Linn. (E. B. t. 525.) P. VII.–IX.

TRIBE X. VELLEÆ.

THALAMIFLORÆ. 15

Tribe XI. RAPHANEÆ.

CRAMBE. KALE.

C. maritima (*sea Kale.*)—Sandy seashores, not common. Plant smooth and branched, about 2 feet high, with a stout, hard stock. Lower leaves roundish, waved and coarsely toothed, yellowish-green, stalked; upper leaves few and much smaller. Flowers in a dense terminal cluster, large and white. Dawlish. Slapton Sands, from whence it was brought for cultivation in 1795, *Fl. D.* (E. B. t. 924.) P. VI.

RAPHANUS. RADISH.

1. **R. Raphanistrum** (*wild R., or jointed Charlock.*)—In cornfields, frequent. Stem from 12 to 18 inches high. Lower leaves lyrate and toothed; upper leaves narrow, toothed, but sometimes entire. Flowers yellow, sometimes reddish, or white, with lilac streaks. Pods·cylindrical, with a long beak. Torquay, etc., common. (E. B. t. 856.) A. VI. VII.

2. **R. maritimus** (*sea Radish.*)—By the seacoast. Rare. Probably merely a seaside variety of the former; growing sometimes as high as 3 or 4 feet. Differs from the last in its irregularly lyrate leaves and larger flowers and seed-pods. Various places around the Bay. (E. B. t. 1643.) A. or B. VI.–VIII.

Ord. VII. **RESEDACEÆ.**

RESEDA. DYER'S-ROCKET, MIGNONETTE.

R. Luteola (*common D., Yellow-weed or Weld.*)—In waste grounds and stony pastures. An erect plant, with a hardy stiff stem, not much branched, about 2 feet high, smooth. Leaves long and lance-shaped, slightly waved at the edges. Stems bearing long stiff racemes of numerous yellowish-green flowers, with prominent stamens. Warberry Hill. Fields between Meadfoot and Hope's Nose. Marychurch. (E. B. t. 320.) A. VI.–VIII.

Ord. VIII. **CISTACEÆ.**

HELIANTHEMUM. ROCK-ROSE.

1. **H. vulgare** (*common R.*)—In dry meadows and stony

thickets. A low, much branched, spreading, shrub-like plant, with a woody stem giving off many ascending flowering branches. Leaves opposite, oblong, green on their upper, but hoary on their under surfaces. Flowers terminal in a raceme of from 6 to 8, bright yellow, with broadly-spreading petals. Kingskerswell. Bradley Wood, near Newton, among rocks, *Cistus Helianthemum*, Linn. (E. B. t. 1321.) P. VII.–IX.

2. **H. polifolium** (*white R.*)—On rocky wastes in limestone districts. Very rare and local. Somewhat like the last, but smaller, less straggling and with smaller leaves, which are hoary on both sides. Flowers large and white. Rocks about Daddyhole Plain, Torquay. Babbicombe Down, on the cliffs near the sea. *Cistus*, Linn. (E. B. t. 1322.) P. VI.–VIII.

Ord. IX. VIOLACEÆ.

VIOLA. VIOLET.

1. **V. palustris** (*Marsh V.*)—In bogs and marshy places. Stock sending out runners; plant usually smooth, with heart-shaped or kidney-shaped leaves, slightly puckered at the edges. Flowers of a pale, delicate blue, streaked with purple, scentless. Bovey Heath. Bogs about Dartmoor. (E. B. t. 444.) P. IV.–VI.

2. **V. hirta** (*hairy V.*)—In woods and pastures, chiefly in limestone districts. Without creeping scions. Leaves heart-shaped, on long stalks, both hairy. Flowers scentless, of a dull blue colour. Hope's Nose. Chudleigh. (E. B. t. 894.) P. IV. V.

3. **V. odorata** (*sweet V.*)—On hedge-banks, in woods and pastures. Stock sending off creeping runners. Leaves coming in a bunch from the crown of the root, heart-shaped and stalked, generally smooth, or slightly downy. Flower-stalks rather longer than those of the leaves, flowers nodding, reddish-purple, or white, very sweet-scented. Cockington lanes. Chelston. Ilsham. (E. B. t. 619.) P. III. IV.

4. **V. canina** (*Gerard's, or Dog V.*)—On banks, in woods and dry pastures. Radical leaves tufted, and flowering branches at first short, but the lateral flowering branches become afterwards much lengthened, rising up from a few inches to sometimes a foot high. Leaves broadly heart-shaped, pointed at the ends; stipules on the flower-stalks narrow lance-shaped and pointed. Flowers often very numerous and large in proportion to the size of the plant, varying in colour from purple to blue, and frequently white, always scentless. Common in fields, wood and pastures around Torquay, etc. (E. B. t. 620, & E. B. S. t. 2736.) P. IV. V.

5. **V. pumila** (*Dillenius's Violet.*)—Boggy heaths, and in sandy places. Plant more upright than in the other species. Leaves longer and much narrower. Flowers either pale blue or yellowish. Buckland. Bovey Heath. *V. Lactea*, Sm. (E. B. t. 445.) P. IV. V.

6. **V. tricolor** (*Pansy, or Heart's-ease.*)—On banks, hilly pastures and cultivated fields. A strong branching plant, with stalked leaves either oval or heart-shaped, but always obtuse; stipules broad and divided into several segments. This plant is however extremely variable in all its parts. Flowers sometimes purple, or yellow, or cream-coloured, sometimes variegated with all three colours. Common on the hilly ground around Torquay, as is also the variety β of Hooker and Arnott. (E. B. t. 1287.) A. V.–IX.

Ord. X. DROSERACEÆ.

DROSERA. SUNDEW.

1. **D. rotundifolia** (*round-leaved Sundew.*)—In bogs and moist heaths, plentiful. Leaves all radical, on long stalks, round, and covered on the upper surface with red sticky hairs, each having a little gland on the top. Flower-stalks slender, rising from amidst the tuft of leaves, and bearing either one or two clusters of pretty little white flowers. Leaves of this, as of all the species of Sundews, more or less covered with small insects, which are entrapped by the viscid juice secreted by the glands of the hairs. Forde bog, near Newton. Bovey Heath. A boggy patch of turf near the road at Spitchwick. Boggy grounds about Ivybridge and Chagford. (E. B. t. 867.) P. VII. VIII.

2. **D. longifolia** (*spathulate-leaved S.*)—In similar situations to the last, from which it is distinguished by its leaves being upright, much longer than they are broad, and being tapered into the leafstalk. Bovey Heath. (E. B. t. 868.) *D. intermedia*, Bab. P. VII. VIII.

Ord. XI. POLYGALACEÆ.

POLYGALA. MILKWORT.

P. vulgaris (*common Milkwort.*)—Hilly and dry pastures, common. Stem herbaceous, procumbent, giving off several either procumbent or ascending branches, bearing scattered linear or oblong leaves; branches from 3 or 4 to 8 inches long. Flowers either blue, purple, pink, or white, the corolla beautifully crested. Daddy-hole Plain, and many places about Torquay and Marychurch. (E. B. t. 76.) P. v.-ix.

Ord. XII. FRANKENIACEÆ.

Ord. XIII. ELATINACEÆ.

Ord. XIV. CARYOPHYLLACEÆ.

Subord. I. *SILENEÆ.*

DIANTHUS. PINK.

D. Armeria (*Deptford Pink.*)—In pastures and on hedgebanks. Stem from 1 to 2 feet high, erect and branching, downy. Leaves linear and slightly hairy. Flowers clustered, rose-coloured, with white dots. Ilsham. Upton Lane, near Torquay. Near Newton, on the Ashburton road. (E. B. t. 317.) A. vii. viii.

SAPONARIA. SOAPWORT.

S. officinalis (*common S.*)—By roadsides, in borders of woods and on hedge-banks. Plant from 1 to 3 feet high, stout and leafy. Leaves broadly lanceolate, opposite. Flowers handsome, of a beautiful rose-colour, in a large terminal cluster. Shaldon. Banks of the Teign, at Teignbridge. (E. B. t. 1060.) P. viii.

SILENE. CATCHFLY.

1. **S. inflata** (*bladder Campion.*)—By roadsides, in fields, and on banks. Stem erect, from 2 to 3 feet high. Leaves somewhat oval, lanceolate. Flowers numerous, drooping, in a terminal panicle, white, the petals deeply cloven; calyx inflated, nearly globular and very much veined. Warberry Hill. Park Hill wood. Meadfoot; and common in fields and waysides. (E. B. t. 164.) P. vi.-viii.

S **maritima** (*sea Campion or Catchfly.*)—Rocky banks

by the seashore. A much smaller plant, with shorter stems and blunter and more fleshy leaves than the last. Flowers larger. Meadfoot Cliffs, abundant. (E. B. t. 957.) P. VI.-VIII.

3. **S. Anglica** (*English C.*)—Sandy and gravelly soils. Plant from 6 inches to a foot high, much branched, slightly viscid. Leaves lanceolate. Flowers single, from the axils of the leaves, white, or tinged with red. Kingsteignton. Dawlish. Lustleigh. (E. B. t. 1178.) A. VI.-X.

LYCHNIS. CAMPION, LYCHNIS.

1. **L. Flos-cuculi** (*meadow L., or Rayged Robin.*)—In moist meadows and pastures. Growing from 1 to 2 feet high. Stem hairy below, clammy above. Leaves lanceolate. Flowers bright rose-colour. Calyx and flower-stalks reddish-purple; petals divided into four segments. Road to Ansti's Cove. Ansti's Cove. Hedge-banks on the side of the Newton road, near Kingskerswell. (E. B. t. 573.) P. V. VI.

2. **L. vespertina** (*white Campion.*)—Under hedges and in bushy places. From 1 to 2 feet high, rather hairy. Leaves ovate-lanceolate. Flowers white, sweet-smelling in the evening. Calyx of the fruitful flowers much more swollen than in the barren ones. Meadfoot Cliffs, and formerly on the Waldon Hill. Meadow between Milber down and the Newton road. (E. B. t. 1580.) B. or P. VI.-IX.

3. **L. diurna** (*red C.*)—Woods and damp hedge-banks, common. Somewhat like the last. Leaves more ovate. Flowers rose-coloured, very rarely white. Torre Abbey fields. Banks and hedges in all the lanes, etc. (E. B. t. 1579.) *L. dioica*, Linn. P. VI.-IX.

AGROSTEMMA. COCKLE

A. Githago (*corn Cockle.*)—In cornfields, very abundant. Stem from 1 to 2 feet high, sometimes slightly branched, covered with long, soft, whitish hairs. Leaves narrow-lanceolate. Flowers large, bright reddish-purple, on long stalks; divisions of the calyx exceedingly long, projecting in long points beyond the petals. Cornfields everywhere. (E. B. t. 741.) *Lychnis Githago*, Bab. A. VI.-VIII.

SUBORD. II. *ALSINEÆ*.

SAGINA. PEARLWORT.

1. **S. procumbens** (*annual small-flowered P.*)—In waste

places, and dry stony pastures, very common. A diminutive plant, with many decumbent branches; seldom more than 2 inches high. Leaves nearly linear. Flowers small and white. Abundant about Torquay and its neighbourhood. (E. B. t. 880.) P. v. IX.

2. **S. apetala** (*annual small-flowered P.*)—In dry gravelly places and on walls. More slender than the last, less branched and with narrower leaves; petals of the flowers very small or wanting. Torquay, etc., abundant. (E. B. t. 881.) P. v.-IX.

3. **S. nodosa** (*knotted P., or Spurrey.*)—In wet, sandy, or marshy places. Plant from 2 to 3 inches high, with many decumbent stems. Lower leaves long and sheathing; stem-leaves shorter, with little tufts of young leaves in their axils. Flowers on short stalks, white, and large in comparison with the size of the plant. Bovey Heath. Gidleigh, near Chagford. *Spergula*, Linn. (E. B. t. 694.) P. VII. VIII.

4. **S. subulata** (*awl-shaped P., or Spurrey*).—Gravelly and stony pastures. Plant small, decumbent at the base, branched, with long solitary flower-stalks. Leaves much like those of *S. procumbens*, but longer. Flowers white. Forde, near Newton Haldon. (*Spergula*, E. B. t. 1082.) P. VI.-VIII.

ARENARIA. SANDWORT.

1. **A. serpyllifolia** (*Thyme-leaved S.*)—On walls, dry banks, and waste places. From 3 to 6 inches high, very much branched, slender and slightly downy. Leaves sessile, very small, acutely ovate. Flower-stalks slender, arising from the forks of the stem. Flowers small and white. Park Hill, Torquay. (E. B. t. 923.) A. VI.-VIII.

2. **A. trinervis** (*three-nerved S.*)—In moist places and shady woods. Plant from 4 or 5 inches to a foot long, tender and much branched, hairy. Leaves oval, with acute ends, stalked, light green, and with 3 distinct nerves. Flowers solitary, white. Wood near Shiphay. *Moehringia*, Bab. (E. B. t. 1483.) A. V. VI.

STELLARIA. STITCHWORT.

1. **S. media** (*common Chickweed or S.*)—In waste places and by roadsides. Stem weak, very variable as to size, much branched, and without hairs, with the exception of an alternate hairy line.

Leaves small, bright green, ovate and pointed, stalked; the uppermost leaves sessile and narrower. Flowers small and white. Seed-vessels oblong. Common everywhere. (E. B. t. 537.) A. III.-IX.

2. **S. holostea** (*greater S.*)—Sides of hedges, woods, and bushy places, common. A straggling plant, with weak stems, from 1 to 2 feet high, not hairy; leaves sessile, very long and lanceolate. Flowers in a forked panicle, very large and white. Lanes about Torquay, etc. (E. B. t. 511.) P. IV.-VI.

3. **S. glauca** (*glaucous marsh S.*)—In wet and marshy places. From 12 to 18 inches high; stem branched, smooth, of a bluish-green colour. Leaves linear-lanceolate, sessile. Flowers solitary, white, not quite so large as those of *S. holostea*, and with narrower segments to the calyx. Moist meadows near Torre Abbey. (E. B. t. 825.) P. V.-VII.

4. **S. graminea** (*lesser S.*)—In dry pastures, meadows, and by sides of hedges, frequent. Stems about a foot long, quadrangular, more slender than the two last. Leaves sessile, linear-lanceolate, pointed. Flowers in loose clusters, small and white; divisions of the calyx 3-ribbed; stamens crowned with red anthers. Hope's Nose. (E. B. t. 803.) P. V.-VIII.

5. **S. uliginosa** (*bog S.*)—In marshes, wet ditches, and by the side of rivulets, frequent. Very variable as to size, from 4 inches to 1 foot; a weak and slender plant, with oblong-lanceolate leaves having a callous tip. Flowers in irregular panicles, small and white; petals very minute. Goodrington. Berry Pomeroy. (E. B. t. 1074.) A. V. VI.

MŒNCHIA. MŒNCHIA.

M. erecta (*upright M.*)—In gravelly and stony pastures. A small plant, from 2 to 5 inches high. Root-leaves stalked, and inclining to spathulate, upper leaves linear-lanceolate, opposite, sharp and stiff. Flowers rather large, few, and white. Meadfoot Cliffs. Bovey. *Sagina*, Linn. (E. B. t. 609.) A. V. VI.

CERASTIUM. MOUSE-EAR CHICKWEED.

1. **C. vulgatum** (*broad-leaved M.*)—In waste or cultivated places, woods and pastures, by roadsides and under hedges. Plant hairy, varying in size from 2 or 3 inches to a foot in height. Stem branching below, each branch again dividing into 2 above. Root-leaves very small and stalked; upper leaves sessile, ovate, pointed

at the ends. Flowers white and small, petals bifid. Common in fields, etc. *C. glomeratum*, Bab. (E. B. t. 789.) A. IV.–IX.

2. **C. viscosum** (*narrow-leaved M.*)—Fields, waste grounds, and on the tops of walls. A larger, coarser, and more spreading plant than the last. Leaves longer and narrower; flowers larger, and gathered together in close clusters, or in a loose forked panicle. Frequent about Torquay. (E. B. t. 790.) A. IV.–IX.

3. **C. semidecandrum** (*little M.*)—In dry places and on wall-tops. Plant from 2 to 6 inches high, downy; leaves broadly ovate, sessile; stem branching at the top into a panicle bearing many small white flowers. Meadfoot Cliffs. Marychurch, Paignton. (E. B. t. 1630.) A. IV. V.

4. **C. pumilum.**—A very viscid variety, classed by Bentham with the last. Daddyhole Plain, on the slope opposite Villa Syracusa.

MALACHIUM. MOUSE-EAR CHICKWEED, WATER STARWORT.

M. aquaticum (*water M. or S.*)—Banks of rivers and ditches. Plant from 1 to 3 feet high, branched and straggling, covered with glandular hairs; leaves large, oval heart-shaped, sessile; the root-leaves stalked. Flowers coming from the forks in the stem, white. The plant has much the appearance of a *Stellaria*. In a ditch by the side of a lane leading to Forde bog from the Newton road. Banks of the Dart, near Totness. Banks of the Teign and Exe. *Cerastium*, Linn. (E. B. t. 538.) P. VII. VIII.

ORD. XV. **LINACEÆ.**

LINUM. FLAX.

1. **L. usitatissimum** (*common F.*)—Found sometimes in cornfields. An elegant-looking plant, from 1 to 2 feet high. Leaves distant and alternate, lanceolate; flowers large, of a beautiful purplish blue, in an imperfect corymb. Exmouth. (E. B. t. 1357.) A. VII.

2. **L. angustifolium** (*narrow-leaved pale F.*)—Sandy and chalky pastures, usually near the sea. From 1 to 2 feet high, irregularly branched; leaves linear-lanceolate; flowers pale blue. Park Hill. Daddyhole Plain. Babbicombe down. (E. B. t. 381.) P. VII.

3. **L. catharticum** (*purging F.*)—Pastures, etc., very frequent. Plant from 2 to 8 inches high; stem branching above. Lower leaves opposite and oblong; upper leaves lanceolate. Flowers small and white, in forked, spreading panicles. Park Hill. Daddyhole Plain. (E. B. t. 382.) A. VI.–VIII.

RADIOLA. FLAX-SEED.

R. Millegrana (*Thyme-leaved F., or All-seed.*)—Moist heaths and boggy soils. A very diminutive plant, from 1 to 2 inches high; stems repeatedly forked; leaves ovate, distant from each other. Flowers mostly terminal and solitary, on short stalks, but some growing also from the axils of the branches. Bovey Heath. Haldon. (*Linum Radiola*, Linn. E. B. t. 983.) A. VII. VIII.

ORD. XVI. MALVACEÆ.

LAVATERA. TREE MALLOW.

L. arborea (*sea Tree Mallow.*)—On insulated sea-rocks in the south and south-west of England. Plant with a somewhat woody stem, varying in height from 3 to 8 feet, branching; leaves downy, divided into seven shallow crenated segments, alternate; flowers large, of a shining purplish rose-colour, with dark purple streaks at the base of the petals. Orestone and Thatcher rocks, off Torquay. (E. B. t. 1841.) B. VII.–IX.

MALVA. MALLOW.

1. **M. sylvestris** (*common Mallow.*)—Waysides and waste grounds, very frequent. Plant growing to 2, 3, or 4 feet high; stem erect. Leaves divided into from 5 to 7 deep lobes, with acute angles. Flowers axillary, in bunches of 3 or 4, deep rose-coloured, with purplish veins. Roadsides and lanes in the vicinity. (E. B. t. 671.) P. VI.–IX.

2. **M. rotundifolia** (*dwarf M.*)—Similar localities to the preceding. Stem branching only from the root, decumbent, 8 or 10 inches long; leaves roundish heart-shaped, with from 5 to 7 shallow lobes; flowers small, rose-purple. Road to Ansti's Cove. Babbicombe road, etc. (E. B. t. 1092.) P. VI.–IX.

3. **M. moschata** (*musk M.*)—In pastures and by roadsides. A much more delicate-looking plant than the two last; from 2 to 3 feet high; the leaves much divided into almost linear segments;

flowers crowded closely together at the extremities of the branches, large, and of a beautiful rose-colour. Aller. Newton road, just beyond Kingskerswell. Bovey Heath. Gidleigh, near Chagford. (E. B. t. 754.) P. VII. VIII.

ORD. XVII. TILIACEÆ.

TILIA. LIME.

T. europæa (*common L., or Linden-tree.*)—In woods over nearly the whole of Europe. A handsome, long-lived tree, growing sometimes as high as 120 feet, but usually not more than half that height. It bears sweet-scented flowers of a pale whitish green. It is said that Linnæus derived his own name from the Swedish *Lin*, our Lime or Linden-tree. Black Head, beyond Hope's Nose, apparently wild. Torre Abbey. Cockington, etc. (E. B. t. 610.) T. VII.

ORD. XVIII. HYPERICACEÆ.

HYPERICUM. ST. JOHN'S-WORT.

1. **H. Androsæmum** (*Tutsan.*)—In bushy places and open woods. A shrubby-looking plant; leaves large, ovate, nearly sessile, and opposite. Flowers bright yellow, in a terminal cyme. Hope's Nose. Ansti's Cove, and formerly on the Waldon Hill. (E. B. t. 1225.) P. VII. VIII.

2. **H. perforatum** (*common perforated St. J.*)—In woods, hedges, and thickets, by roadsides, etc., frequent. Plant from 12 to 18 inches high, branching at the upper part; leaves oblong, sessile, marked with numerous pellucid dots; flowers in a showy terminal corymb, bright yellow, marked with little black dots. Waldon Hill. Park Hill. Ansti's Cove, etc. (E. B. t. 295.) P. VII.-IX.

3. **H. dubium** (*imperforate St. J.*)—Similar situations to the last, attached especially to hilly districts. Somewhat like *H. perforatum*, but more leafy and with a slightly quadrangular stem. Leaves larger and broader with fewer clear dots, but with some black ones on their under sides. Berry Pomeroy woods. Many parts of Dartmoor. (E. B. t. 296.) P. VII. VIII.

4. **H. quadrangulum** (*square-stalked St. J.*)—In moist pastures, by hedges, ditches, and rivulets. Readily distinguished from the two foregoing by its square stem with 4 well-marked angles. Leaves clasping, opposite and ovate, flowering stems

arising from the axils; flowers small, yellow, not so showy as the last. Torquay. Marychurch. Cockington lanes. Ilsham. Paignton. Bradley Woods. (E. B. t. 370.) P. VII.

5. **H. humifusum** (*trailing St. J.*)—In stony heaths, and boggy pastures, woods, and thickets. A low, decumbent, very much branched, trailing plant, with small oblong leaves, and not very numerous pale-yellow flowers. Park Hill wood. Meadfoot. Marychurch. Moss-grown blocks of rock on the banks of the Erme, Ivybridge. (E. B. t. 1226.) P. VII.

6. **H. linariifolium** (*linear-leaved St. J.*)—On hilly wastes and in rocky situations. Resembling *H. humifusum*, but more upright in its growth and with much narrower leaves. Flowers in terminal corymbs, larger and of a brighter yellow. Banks of the Teign. Belmont, near Exeter (Miss Snow). Between Sandy Park and Fingle Bridge (G. W. Warren). (E. B. S. t. 2851.) P. VII. VIII.

7. **H. pulchrum** (*small upright St. J.*)—Dry woods, open heaths, and waysides, frequent. Plant upright, with a slender but stiff stem, from 1 to 2 feet high. Leaves of the principal stem somewhat heart-shaped, clasping; those of the branches much smaller and narrower. Flowers in loose clusters, bright yellow; the buds before opening tipped with vivid red. Very common about the neighbourhood of Torquay. (E. B. t. 1227.) P. VI. VII.

8. **H. hirsutum** (*hairy St. J.*)—In copses and woods, frequent. Plant about 2 feet high, with a downy or hairy stem; leaves large, rather downy beneath, oblong, slightly stalked, and marked with numerous pellucid dots. Flowers yellow, but much paler than the last. Park Hill. Meadfoot Cliffs. Bradley Woods. Copse by the brook at Chudleigh. (E. B. t. 1156.) P. VII. VIII.

9. **H. montanum** (*mountain St. J.*)—In woods and bushy places in hilly districts, not so frequent as the other species. About 2 feet high; stem stiff and upright, not branched. Lower leaves large, of a pointed oval shape, clasping the stem, opposite; upper leaves smaller, without shining dots, but having a row of black ones round their under margins. Flowers yellow, in a dense terminal cluster. Ansti's Cove. Babbicombe Down. Milber Down. Chudleigh. (E. B. t. 371.) P. VII. VIII.

10. **H. Elodes** (*marsh St. J.*)—In marshy and boggy places. Stems creeping and spreading, from 6 to 8 or 10 inches long; leaves roundish, opposite, clasping the stem; whole plant covered with loose, woolly, whitish hairs. Flowers yellow, in a loose terminal cluster. Osier-beds at Paignton. Forde bog, near Newton. Bovey Heath. In boggy places near Ivybridge. Chagford. Bogs about Dartmoor. (E. B. t. 109.) P. VII. VIII.

Ord. XIX. ACERACEÆ.

ACER. MAPLE.

1. **A. Pseudoplatanus** (*greater M., or Sycamore.*)—Abundant in the south of England, in plantations and hedges. A large and handsome tree; leaves 5-lobed and unequally serrate; flowers in loose, hanging racemes. The wood is much used by turners for making bowls and trenchers. Common about Torquay. Berry Pomeroy woods, etc. (E. B. t. 303.) T. v. vi.

2. **A. campestre** (*common M.*)—In woods and thickets. A not very tall tree, with corky, fissured bark, and dense dark green foliage; leaves 5-lobed, segments entire, or sometimes slightly cut. Flowers few, on slender stalks, in loose upright corymbs. The wood of this tree is often beautifully veined, and is then much prized. Hedges about Torquay. Cockington. Shiphay. Berry Pomeroy. (E. B. t. 304.) T. v. vi.

Ord. XX. GERANIACEÆ.

GERANIUM. CRANE'S-BILL.

1. **G. lucidum** (*shining Crane's-bill.*)—In stony and waste grounds, on old walls, etc. Stems spreading; leaves 5-lobed and roundish, lobes 3 times cut; both leaves and stem shining; the root-leaves kidney-shaped, often of a bright red. Flowers small and rose-coloured. Common in lanes and on hedge-banks. (E. B. t. 75.) A. v.-viii.

2. **G. Robertianum** (*stinking C., or Herb-Robert.*)—In thickets, woods, waste ground, and by waysides, very common. An upright or spreading, much-branched plant, from 6 to 12 or 14 inches high, slightly hairy. Leaves divided into three pinnate segments. Flowers small, reddish-purple, sometimes white. The whole plant emits a disagreeable smell when rubbed. Common everywhere. A small variety, the *G. purpureum* of Mill, grows on the Rock Walk and cliffs beyond Meadfoot. (E. B. t. 1486.) A. v.-ix.

3. **G. molle** (*dove's-foot C.*)—In dry pastures and waste grounds, common. A spreading, weak-looking plant, usually covered with long, soft hairs. Root-leaves numerous, on long stalks, roundish, divided into from 7 to 11 lobes, which are again cut into 3 or 5 segments; upper leaves smaller and with fewer divisions. Flowers in twos, on short stalks, small, pinkish-purple.

Lanes, fields, and hedge-banks, abundant. (E. B. t. 778.) A. IV.–VIII.

4. **G. rotundifolium** (*round-leared C.*)—Pastures and waste places, not so common as the preceding. Somewhat like the last in appearance, but stouter; leaves not so much divided, and with broader lobes. Flowers smaller and on shorter stalks, petals entire, flesh-coloured. Meadfoot Cliffs. Dartmouth, near the Castle. (E. B. t. 157.) A. V. VI.

5. **G. pusillum** (*small-flowered C.*)—By sides of hedges and in waste ground. Stem weak; leaves rounded or reniform, with from 5 to 7 deep 3-cut lobes. Flowers very small, bluish-purple; petals notched. Ansti's Cove Lane. Babbicombe. (E. B. t. 385.) A. VI.–IX.

6. **G. dissectum** (*jagged-leaved C.*)—In dry pastures, waste and cultivated places: characterized by the very much divided leaves, which are composed of from 5 to 7 or 9 narrow segments; the footstalks of the flowers are also very short, and bear two small purple flowers. Very common. (E. B. t. 753.) A. V.–VIII.

7. **G. columbinum** (*long-stalked C.*)—In dry pastures, on banks, and in waste places. A slender, decumbent plant; leaves deeply divided, with segments still narrower than in *G. dissectum*. Flowers small, rose-coloured, on long, slender stalks; stem in this, as well as in the last, furnished with reflexed hairs. Very frequent about the neighbourhood of Torquay. (E. B. t. 259.) A. VI. VII.

ERODIUM. STORK'S-BILL.

1. **E. cicutarium** (*Hemlock Stork's-bill.*)—In waste and cultivated land, and stony pastures near the sea. Whole plant hairy; stems generally short, but at times attaining as much as 6 or 9 inches in length; nearly all the leaves radical, on long stalks, pinnate, with deeply divided segments. Flower-stalk long, surmounted by an umbel of from 2 or 3 to 12 small purplish or pink flowers. Cliff Walk at Meadfoot. Side of the rocky bank on the right of the pathway leading to Ansti's Cove. (E. B. t. 1768.) A. VI.–IX.

2. **E. moschatum** (*musky S.*)—In sandy waste places and heaths, in the neighbourhood of the sea. Larger and much coarser than the last; often as much as a foot long. Leaves pinnate, with sessile leaflets, deeply toothed. Flowers rather large, and numerous in the umbel, bluish-purple. Whole plant yields a strong musky smell. Paignton green. Teignmouth. Dawlish. (E. B. t. 902.) A. VI. VII.

3. **E. maritimum** (*sea S.*)—Sandy seacoasts, rather rare.

A small, softly hairy plant, distinguished from the other two by its leaves, which are ovate heart-shaped, crenated and stalked. Flower-stalks bearing one or two small reddish-purple flowers; beak of the seed much shorter than in *E. moschatum*. Paignton sands. Teignmouth. (E. B. t. 646.) P. v.–ix.

Ord. XXI. BALSAMINACEÆ.

Ord. XXII. OXALIDACEÆ.

OXALIS. WOOD-SORREL.

1. **O. Acetosella** (*common W.*)—In woods and shady places, frequent. A delicate and beautiful little plant. Leaves all growing from the root on long stalks, with 3 inversely heart-shaped leaflets. Flower-stalks also radical, bearing a single, rather large, white flower with purplish streaks. The leaves have a slightly acid taste. The original of the Irish Shamrock, though now replaced by the far less beautiful Dutch Clover. Cockington lanes. Berry Pomeroy woods. Woods about Gidleigh, near Chagford. Copses on the banks of the Erme, near Ivybridge. (E. B. t. 762.) P. v.

2. **O. corniculata** (*yellow procumbent W.*)—In shady waste grounds. Stem spreading, with decumbent branches; leaves of 3 inversely heart-shaped leaflets. Flower-stalks axillary, slender, carrying an umbel of from 2 to 5 small pale-yellow flowers. Teignmouth. Exmouth Warren. (E. B. t. 1726.) A. vi–ix.

Ord. XXIII. STAPHYLEACEÆ.

Sub-Class II. **CALYCIFLORÆ** (Ord. XXIV.–XLIX.)

Ord. XXIV. CELASTRACEÆ.

EUONYMUS. SPINDLE-TREE.

E. Europæus (*common S.*)—In woods and hedges. A shrub, standing from 3 to 6 feet high, with green and smooth bark, and four-angled branches. Leaves somewhat oval and lanceolate, finely serrated, not hairy; flowers small and white, in a few-flowered umbel; fruit square, with blunt angles, of a beautiful rose-colour. The wood is used for making skewers and spigots. Waldon Hill. Bushy places about Ansti's Cove. Marychurch. Chudleigh. (E. B. t. 362.) Sh. v. vi.

Ord. XXV. RHAMNACEÆ.

RHAMNUS. BUCKTHORN.

R. Frangula (*Alder Buckthorn.*)—In woods and thickets more frequent in England than *R. catharticus*. A small shrub; branches without spines; leaves elliptical, narrower towards the stalk; flowers in small clusters, greenish-white, petals very small; berries dark purple, about the size of a small pea, containing 2 seeds. Sandy Park, near Chagford. Kingsteignton. Heywood, near Exmouth (Rev. J. Jervis); Exwick wood (Mr. Jacob); Widdecomb-in-the-Moor, *Fl. D.* (E. B. t. 250.) Sh. v. vi.

Ord. XXVI. LEGUMINOSÆ.

Tribe I. *GENISTEÆ.*

ULEX. FURZE.

1. **U. Europæus** (*common F., Whin, or Gorse.*)—On heaths, and sandy or stony wastes, in England, Ireland, and the southern parts of Scotland. A strong, hardy shrub, from 3 to 6 feet high, with close-set, spreading branches, armed with sharp, branching spines; young leaves shaggy. Flowers bright yellow, rising from the primary and secondary spines; calyx coarsely hairy. In Devonshire this plant continues flowering during the whole year. Very abundant. Warberry Hill. Meadfoot Cliffs, etc. (E. B. t. 742.) Sh. i.-xii.

2. **U. nanus** (*dwarf F.*)—On dry heaths, in many places in England. More strictly western than *U. Europæus*. It is much smaller than the last in all its parts. Stem procumbent; calyx covered with a fine down; flowers not more than half the size of *U. Europæus*. The spines short and spreading, branched at their base only. Thought by some botanists to be merely a variety; but its characteristics are so constant, that it certainly seems entitled to be considered as a distinct species. Bovey Tracey, and on exposed downs in many parts. (E. B. t. 743.) Sh. vii.-xi.

GENISTA. GREEN-WEED.

G. Anglica (*needle G., or Petty-whin.*)—In wet heaths and moory ground. A small, loosely-branched, spreading plant, scarcely a foot high, with ovate-lanceolate leaves, and bearing simple spines; flowering branches spineless; flowers yellow, in short axillary racemes, accompanied by leaf-like bracts; pods about half an inch long, broad and swelling. Bovey Heath. (E. B. t. 132.) Sh. v. vi.

SAROTHAMNUS. BROOM.

S. scoparius (*common Broom.*)—On high hilly wastes and bushy places. Plant from 2 or 3 to 6 feet high, with numerous long, upright, wiry branches, which are green and prominently angled. Leaves ternate, stalked; leaflets obovate; upper leaves simple and sessile. Flowers large, golden-yellow, single or in pairs, borne on slender stalks in the axils of the old leaves. Pod hairy at the sides, many-seeded. Cliffs at Meadfoot. Walks at Ilsham. Downs above the Teign, near Whyddon Park. (*Spartium*, E. B. t. 1339.) Sh. v. vi.

ONONIS. REST-HARROW.

O. arvensis (*common Rest-harrow.*)—In barren pastures, ill-cultivated lands, and borders of fields; very variable in its appearance, sometimes erect, but usually procumbent and rooting; stem generally clothed with soft hairs, and having a sticky feel to the touch, sometimes spinous. Leaves oval or oblong, somewhat serrate: lower leaves ternate. Flowers on short stalks, solitary, sometimes white, but generally rose-coloured, pod 2- or 3-seeded. Very frequent. Fields on the Warberry Hill. Walks above Meadfoot. Paignton Green, etc. *O. campestris*, Bab. (E. B. t. 682.) P. vi.-ix.

ANTHYLLIS. KIDNEY-VETCH.

A. Vulneraria (*common K., or Ladies'-fingers.*)—In dry pastures, and rocky or stony places in hilly districts. Stems spreading or ascending from 3 or 4 inches to a foot long. Leaves pinnate, with from 5 to 9 lanceolate, hairy leaflets, the terminal one the longest. Flower-heads in pairs at the ends of the branches, with large palmated bracts; flowers small, and varying in colour from yellow to deep red, calyx hairy; pod from 1- to 3-seeded. Meadfoot Cliffs. Daddyhole Plain. Petit Tor. Watcombe. (E. B. t. 104.) P. VI.–VIII.

Tribe II. *TRIFOLIEÆ*.

MEDICAGO. MEDICK.

1. **M. lupulina** (*black Medick, or Nonsuch.*)—In waste places and pastures, abundant. Stems spreading, from 1 to 2 feet long, more or less hairy; leaflets obovate; flower-stalks long; flowers small, numerous, yellow, in dense oval spikes. Pods small, kidney-shaped, turning black when ripe, scarcely spiral, 1-seeded. Meadfoot. Ilsham. (E. B. t. 971.) A. V.–VIII.

2. **M. maculata** (*spotted M.*)—In cultivated and waste places. Very much, in habit, like the last, but having obcordate leaflets marked in their centre with a purple spot. Flower-stalks from 1- to 4-flowered; pods compressed, and making 2 or 3 spiral turns, armed with 2 rows of spreading, curved prickles. Banks at Meadfoot. Ilsham. Hope's Nose. (*M. polymorpha*, E. B. t. 1616.) A. V.–VIII.

3. **M. minima** (*little Bur-Medick.*)—In open pastures and waste places, rare. Like the last, but more compact and smaller, softly hairy or downy; flower-stalks from 1- to 6-flowered. Legumes smaller and nearly round, with 2, 3, or 4 close spiral turns, and edged with a double row of hooked spines. Banks by the cliff-walks between Meadfoot and Hope's Nose. (E. B. S. t. 2635.) A. V.

MELILOTUS. MELILOT.

M. officinalis (*common yellow Melilot.*)—By roadsides, on banks, and in bushy places. Plant generally erect, branched, from 2 to 4 feet high; leaves on long leafstalks, distant; lower leaves with roundish leaflets; leaflets of the upper leaves nearly linear. Flowers in loose lateral racemes, bright yellow; pods ovate, pointed, wrinkled, and hairy. Babbicombe. Lanes about Barton. Watcombe. *Trifolium*, Sm. (E. B. t. 1340.) A. or B. VI.–VIII.

TRIGONELLA. FENUGREEK.

T. ornithopodioides (*bird's-foot Fenugreek.*)—In dry, sandy pastures; stems prostrate and spreading, 2 or 3 inches long; leaflets congregated at the summit of the stalk, obcordate, toothed. Clusters axillary, stalked, of from 1 to 3 small, nearly white flowers; pod curved, compressed, transversely furrowed, 6- to 8-seeded. Babington says, this plant is scarcely a *Trifolium* or *Trigonella.* Paignton sands. *Trifolium*, Linn. (E. B. t. 1047.) A. VI. VII.

TRIFOLIUM. CLOVER, TREFOIL.

1. **T. repens** (*white Trefoil, or Dutch Clover.*)—In meadows and pastures. Stems creeping and rooting at the joints; leaflets obovate or obcordate, with usually a dark spot at their base. Flower-stalks long and erect, bearing a roundish head or umbel of white flowers, often pinkish. Pod protruding beyond the calyx, and covered by the withered corolla; seeds from 2 to 4. Very common in pastures, etc. (E. B. t. 1769.) P. V.-IX.

2. **T. pratense** (*common purple C.*)—Frequent in meadows and pastures. Stems hairy from 1 to 2 feet long; leaflets oval, or obcordate; flowers in dense ovate sessile heads, reddish-purple, sometimes white; pod generally 1-seeded; flower-heads with 2 sessile trefoiled leaves at their base. Abundant everywhere. (E. B. t. 1770.) P. V.-IX.

3. **T. medium** (*zigzag T.*)—In open woods, bushy pastures, banks, and waysides. Differing from the last in its zigzag stem and stalked flower-heads; it has also narrower stipules and leaflets. Flowers in large heads, purplish. Woods and banks near Torquay. Near Ide on the Moreton road. (E. B. t. 190.) P. VI.-IX.

4. **T. incarnatum** (*crimson C.*)—In open places near the sea. An erect, slender, and softly hairy plant, naturalized in a few places in the south of England; leaflets obcordate; flower-heads ovate or cylindrical; flowers rich crimson. Meadows about Torquay, but probably cultivated. The variety β of Hooker and Arnott, with yellow flowers, is said to be "decidedly indigenous" at Lizard Point, Cornwall. (E. B. S. t. 2950.) A. VI. VII.

5. **T. arvense** (*hare's-foot T.*)—In cornfields, dry pastures, and on sandy banks. Stem from 6 to 12 inches high, erect, and much branched; leaflets lanceolate, blunt; flower-heads numerous, cylindrical, soft and hairy; flowers very small, pale pinkish-white; the projecting hairy teeth of the calyx give the heads a peculiar

feathery appearance. Meadfoot Cliffs. Hope's Nose. (E. B. t. 944.) A. VII.–IX.

6. **T. striatum** (*soft-knotted T.*)—In dry pastures and waste places. Whole plant softly hairy; stem from 4 to 10 inches long, more or less spreading; leaflets obovate. Flower-heads terminal and axillary, small, oval or globular; flowers very small, pale red; calyx with 5 short, awl-shaped teeth. Paignton Green. (E. B. t. 1813.) A. VI. VII.

7. **T. scabrum** (*rough rigid T.*)—In sandy pastures and waste places near the sea. Somewhat like the last, but more prostrate and not so hairy; leaflets not so broad, and the flower-heads more in the axils of the leaves, and very rigid when in fruit; flowers very pale. Paignton sands. Kerswell down. (E. B. t. 903.) A. V.–VII.

8. **T. glomeratum** (*smooth round-headed T.*)—On dry heaths, in pastures and waste grounds. A small, slender, spreading plant, very much like *T. scabrum*, but with rounder heads and flowers of a much brighter pink; corolla very small, yet longer than the calyx-teeth. Ellacombe. Ilsham Down. Exmouth sands. (E. B. t. 1063.) A. VI.

9. **T. suffocatum** (*suffocated T.*)—Sandy seashores, rare. A very small plant with prostrate stems, seldom more than 2 inches long; leaflets obovate, on slender footstalks. Flowers very small, closely sessile, in minute dense heads, which are crowded along the slender stems quite close to the ground. Mound at Ellacombe. Paignton Green. Teignmouth Den. (E. B. t. 1049.) A. VI.

10. **T. subterraneum** (*subterranean T.*)—In dry, gravelly or sandy pastures. Stem from 3 to 8 inches long, prostrate, hairy; leaflets obovate on long leafstalks. Flowers almost white, long, and very slender, in heads of 2 or 3, on axillary, lateral stalks, which after flowering become elongated and turn downwards, and by means of the 5 spreading teeth, which arise and surround the calyx, the fruit is buried in the ground. Ellacombe. Chapel Hill. Paignton Green. Kerswell down, Teignmouth (Miss Champernowne). (E. B. t. 1048.) A. V. VI.

11. **T. fragiferum** (*Strawberry-headed T.*)—In pastures and meadows. Plant in habit and appearance resembling *T. repens*, but differing in the flower-heads, which consist of numerous closely sessile flowers, surrounded by a much divided involucre, and the calyx, after flowering, becomes dilated and coloured, so that the fruiting head has much the appearance of a Strawberry. Torre Abbey meadows. Goodrington Marsh. (E. B. t. 1050.) P. VII. VIII.

12. **T. procumbens** (*Hop T.*)—In dry pastures and borders

of fields. Plant slender, from 6 inches to a foot long, either procumbent or erect. Leaflets obovate or obcordate, the middle one distant from the others. Flower-heads broadly oval, dense and many-flowered; flowers small and yellow, standard becoming striated as the flowers fade. The whole head has a Hop-like appearance. Frequent in pastures. Torquay. Marychurch, etc. (E. B. t. 945.) A. VI.-VIII.

13. **T. minus** (*lesser yellow T.*)—By roadsides and in dry pastures. More slender and procumbent than the last; flowers not so numerous in the heads, and paler; the standard narrower, and more faintly striated. Ilsham. Marychurch, etc., frequent. (E. B. t. 1256.) A. VI.-VIII.

14. **T. filiforme** (*slender yellow T.*)—In sandy or stony pastures and in waste places near the sea. Stems seldom attaining 6 inches in length; leaflets narrow, and the centre one placed immediately between the two others. Flowers from 2 or 3 to 5 or 6 in a head, very small. Starved specimens of *T. minus* are very often mistaken for this. According to Bentham, "rare in Britain," though set down as "frequent" by Hooker and Arnott. Park Hill. Mound at Ellacombe. Ilsham Down. King's Kerswell, on the rocky mound near the Church. (E. B. t. 1257.) A. VI. VII.

LOTUS. BIRD'S-FOOT TREFOIL.

1. **L. corniculatus** (*common B.*)—In meadows and pastures. Plant varying in size from 2 inches to nearly 2 feet long; stems decumbent or erect; leaves ovate and pointed. Flower-heads in the form of an umbel, from 5- to 10-flowered, flower-stalks long; flowers bright yellow, the outside of the standard frequently red. Pods about 1 inch long, containing globular seeds. Daddyhole Plain. Meadfoot. Babbicombe. The variety β *villosus*; rocks near Hope's Nose. (E. B. t. 2090.) P. VI.-VIII.

2. **L. major** (*narrow-leaved B.*)—In moist meadows and along the sides of ditches. More generally luxuriant in its whole growth than the last, of which it is most probably a highly developed form. Heads with from 6 to 12 flowers in the umbel. Road to Ansti's Cove. Marychurch. Walks at Meadfoot. (E. B. t. 2091.) P. VI.-VIII.

3. **L. angustissimus** (*slender B.*)—In meadows, pastures, and fields. Rare. A slender, branched, and hairy plant, with narrower leaflets than *L. corniculatus*. Flowers small, solitary, or 2 or three in the umbel. Pod slender, about ¼ an inch long. Meadfoot. Petit Tor. Roadside between Lindridge and Bishopsteignton. (*L. diffusus*, E. B. t. 925.) A. V.-VIII.

4. **L. hispidus** (*hispid B.*)—A larger and more hairy variety of the last. Flowers often 3 in the umbel; pod thicker and shorter. Hedgerow by the fir-plantation near Hope's Nose (Mr. C. Parker). (E. B. S. t. 2823.) A. v.–viii.

TRIBE III. ASTRAGALEÆ.

TRIBE IV. HEDYSAREÆ.

ORNITHOPUS. BIRD'S-FOOT.

O. perpusillus (*common B.*)—In sandy and dry rocky pastures. Stems spreading on the ground, from 2 or 3 to 8 inches long. Leaves pinnate, with from 5 to 10 pairs of oval or oblong leaflets, with an odd terminal one, softly hairy. Flower-stalks longer than the leaves; flowers white with red lines, nearly sessile. Pods downy, curved upwards, terminating in a beak. Daddyhole Plain. Top of the cliffs, near Hope's Nose. Babbicombe and Ilsham Downs. Bovey Heath. Chagford. (E. B. t. 869.) A. v.–vii.

HIPPOCREPIS. HORSE-SHOE VETCH.

H. comosa (*tufted Horse-shoe Vetch.*)—In pastures and on banks, in limestone districts chiefly. Stems numerous, branching at the base; leaves pinnate, leaflets small, oblong or linear, from 9 to 15. Flowers umbellate, from 5 to 8, pale yellow, resembling those of *Lotus corniculatus*, the pods however are quite different, and distinguished by the broad and deep semicircular notches at their inner edges. Daddyhole Plain. Ansti's Cove. Babbicombe Down. (E. B. t. 31.) P. v.–viii.

ONOBRYCHIS. SAINFOIN.

O. sativa (*common Sainfoin.*)—On limestone hills, and dry open downs. Stems ascending, from 1 to 2 feet high; leaves pinnate, with numerous oblong leaflets. Flower-stalks long, bearing a spike of light pink flowers; pods sessile, bordered with short teeth, the sides marked with elevated veins. Fields above the walks from Meadfoot to Hope's Nose. *Hedysarum Onobrychis*, Linn. (E. B. t. 96.) P. v.–viii.

Tribe V. Vicieæ.

VICIA. VETCH, TARE.

1. **V. lathyroides** (*spring Vetch.*)—In dry pastures, open woods, and on hedge-banks. A low spreading plant; stems from 3 to 6 inches long, branching below; leaflets from 2 to 6, oblong or narrow-linear. Flowers sessile, solitary, rich purple; distinguished from the next chiefly by its smaller flowers, and by its seeds being rough with raised dots. Ansti's Cove. Hedge-banks on the Teignmouth road. Side of the Erme, near Ivybridge. (E. B. t. 30.) A. v.–vi.

2. **V. sativa** (*common Vetch.*)—In dry pastures and cultivated ground. Stems from 1 to 2 feet high, nearly erect, spreading or nearly climbing; leaflets varying from elliptic-oblong or obcordate to narrow-linear, 4 to 7 on each leaf. Flowers usually in pairs, large, purplish-blue or red. Hedges and fields, frequent, probably cultivated. (E. B. t. 334.) The variety *V. angustifolia*, with narrower leaflets and smaller flowers (E. B. S. t. 2614), is to be found by the Cliff walks at Ilsham. A. v. vi.

3. **V. sepium** (*bush V.*)—In woods and shady places. Stem from 1 to 2 feet high; leaflets large, ovate, from 4 to 8 pairs to each leaf, ending in a tendril; flowers 2, 4, or 6 together, in the axils of the upper leaves, pale reddish-purple, forming a drooping cluster on a short flower-stalk. Legume about an inch long. Common in woods and hedges. (E. B. t. 1515.) P. vi. vii.

4. **V. lutea** (*rough-podded yellow V.*)—In dry, stony, waste or cultivated land. A slightly hairy plant, with spreading, branched stems from 6 to 12 inches long; leaves with elliptic-lanceolate leaflets, 6 to 9 pairs on a leafstalk, varying greatly in hairiness. Flowers large, pale yellow; pods compressed and clothed with long hairs. Near Hope's Nose. (E. B. t. 481.) P. vi.–viii.

5. **V. Bithynica** (*rough-podded purple V.*)—In bushy places or stony wastes near the sea. Rare. Plant generally slightly downy; stems from 1 to 2 feet long, weak and angular. Leaves with generally but 2 pairs of leaflets varying much in breadth, sometimes long-lanceolate, sometimes nearly linear; tendrils branched. Flowers solitary or in twos, on shorter or longer flower-stalks, purple, with whitish wings. Pods upright, rough, from 1 inch to an inch and half long, containing from 4 to 6 seeds. Maidencombe (Miss A. Griffiths). (Teignmouth. Shaldon, *Fl. D.*) (E. B. t. 1842.) P. vii. viii.

6. **V. Cracca** (*tufted V.*)—In hedges and bushy places. Stems 2 or 3 feet long, weak and climbing; leaflets lanceolate or linear,

slightly hairy. Flowers numerous, in one-sided clusters, of a beautiful bluish-purple; legume about an inch long, flattened, 6- to 8-seeded. Warberry Hill. Hedges by the Newton road, etc. Common. (E. B. t. 1168.) P. VI.–VIII.

7. **V. sylvatica** (*wood V.*)—Open woods and bushy places in rocky countries. A very handsome plant, spreading its stems over bushes and small trees to the extent of 8 or 10 feet. Leaves bearing from 8 to 10 pairs of ovate or oblong leaflets; flower-stalks longer than the leaves; flowers numerous, extremely beautiful, in long drooping clusters, white, streaked with purplish veins. Pod about an inch long, broad, containing from 4 to 6 seeds. Ansti's Cove. Petit Tor. Bushy places near Hope's Nose. (E. B. t. 79.) P. VI.–VIII.

8. **V. tetrasperma** (*slender Tare.*)—In fields, hedges, and waste places. Stems weak and spreading, usually 6 or 8 inches long, but sometimes climbing to the extent of 2 feet. Leaves with from 3 to 6 pairs of linear-obtuse or acute leaflets. Flower-stalks slender, bearing from 2 to 7 small, pale-blue flowers. Pods linear-oblong, flat, containing from 4 to 6 seeds. Warberry Hill. Cliffs at Meadfoot. *Ervum*, Linn. (E.B. t. 1223.) A. VI.–VIII.

9. **V. hirsuta** (*hairy Tare.*)—In hedges, cornfields, and waste places, frequent. Stems from 6 inches to 2 or 3 feet long, weak and straggling. Leaves with from 6 to 8 pairs of small oblong leaflets; flower-stalks slender, bearing two or three very small pale blue flowers; teeth of the calyx longer than the tube of the corolla. Legume about ¼ inch long, flat and hairy, 2-seeded. Warberry Hill. Fields above Meadfoot. Walks near Hope's Nose. *Ervum*, Linn. (E. B. t. 970.) A. VI.–VIII.

LATHYRUS. VETCHLING, EVERLASTING PEA.

1. **L. Nissolia** (*crimson V. or Grass Vetch.*)—In grassy borders of fields, bushy places, or stony pastures, rare. Stem about a foot high, branching from the bottom, without true leaves, but the leafstalk flattened out into a linear-lanceolate expansion resembling a blade of grass. Flower-stalks long, with 1 or sometimes 2 small, reddish flowers. Ellacombe (Miss A. Griffiths). (Teignmouth. Exmouth, *Fl. D.*) (E. B. t. 112.) A. VI.

2. **L. pratensis** (*meadow V.*)—In moist meadows and pastures. A weak, much branched, straggling or climbing plant, attaining the length of 2 or 3 feet; tendrils branched, with 2 lanceolate, 3-nerved leaflets; stipules large and arrow-shaped. Flower-stalks long, with a cluster of numerous yellow flowers. Ansti's Cove. Walks above Meadfoot, etc. (E. B. t. 670.) P. VII. VIII.

3. **L. sylvestris** (*narrow-leaved Everlasting Pea.*)—In hedges, thickets, and bushy, rocky places. Stems straggling and climbing, from 2 to 5 or 6 feet long, broadly winged; leafstalks with 1 pair of ensiform leaflets and ending in a tendril. Flower-stalks long, many-flowered; flowers greenish, streaked with purple veins. Legumes 2 or 3 inches long, containing numerous flattened seeds. Meadfoot Cliffs. Hope's Nose. Ansti's Cove. Maidencombe. Berry Head, etc. (E. B. t. 805.) P. VII.–IX.

OROBUS. BITTER VETCH.

O. tuberosus (*tuberous Bitter Vetch.*)—In thickets and open woods, under hedges, etc. Root forming small tubers; stems suberect, from 6 inches to a foot long; leaves with from 2 to 4 pairs of leaflets, without tendrils, oblong-lanceolate or linear. Flowers reddish-purple, from 2 to 4, on slender flower-stalks. Cliff walks near Hope's Nose. Ansti's Cove. (*Lathyrus macrorrhizus*, Bab. and Benth.) (E. B. t. 1153.) The variety β, with linear leaflets (*Orobus tenuifolius*), is found at Bovey Tracey. P. VI. VII.

Ord. XXVII. ROSACEÆ.

Subord. I. *AMYGDALEÆ*.

PRUNUS. PLUM or CHERRY.

1. **P. communis** (*common Plum.*)—Abundant in hedges, thickets, and open woods. A much-branched spiny shrub, the smaller branches ending in a sharp thorn; leaves stalked, ovate or oblong; flowers small and white, appearing before the leaves. *P. spinosa*, Linn. (E. B. t. 842.) Var. β. *insititia:* less spinous branches, more downy leaves, and of taller growth. (E. B. t. 841.) Var. γ. *domestica:* branches without spines; fruit larger and sweeter. (E. B. t. 1783.) The two first grow abundantly about the neighbourhood; the last about Cockington. Walks near Hope's Nose. Sh. IV. V.

2. **P. Avium** (*wild Cherry or Gean.*)—In woods and hedges. A tree, from 20 to 30 or more feet high, gracefully branching; leaves drooping, oblong-ovate or ovate-lanceolate, toothed. Flowers in clusters of 2 or 3, white, rising together from leafless buds. Fruit roundish, somewhat heart-shaped, firm, either black or red. The original of the common garden Cherry. Shiphay lanes. Cockington lanes. Bradley woods. Berry Pomeroy woods. Chudleigh. (*P. Cerasus*, Sm. in E. B. t. 706.) T. V.

3. P. Cerasus (*Morello Cherry.*)—In woods, hedges, and thickets. An upright, bushy shrub, from 6 to 8 feet high; leaves oblong-obovate or broadly ovate-lanceolate, *not drooping;* flowers white, in scattered umbels. Fruit always red, juicy and acid, round. Berry Pomeroy woods. Copse at Chudleigh. (E. B. S. t. 2863.) Sh. v.

SUBORD. II. *ROSEÆ.*

TRIBE I. SPIRÆIDÆ.

SPIRÆA. DROPWORT OR MEADOW-SWEET.

1. S. Filipendula (*common Dropwort.*)—In meadows, pastures, and open woods, very frequent. A beautiful little plant; stems erect, from 1 to 2 feet high; leaves confined to the lower part of the stem, mostly radical, interruptedly pinnate, with numerous small oblong or lanceolate segments, deeply cut and serrate. Flowers small, cream-coloured, tipped with pink, in a compound cluster terminating the stem. Plentiful in meadows and pastures about Torquay and Marychurch. (E. B. t. 284.) P. VI. VII.

2. S. Ulmaria (*Meadow-sweet, Queen of the Meadows.*)—In meadows and by the sides of ditches and ponds. Plant from 2 to 4 feet high; stem branched and furrowed. Leaves large, pinnate, with ovate or broadly lanceolate segments, green above, whitish beneath; terminal segment the largest and divided into 3 lobes. Flowers sweet-scented, small, yellowish-white, in compound cymes. Fruit twisted spirally. Meadows behind Torre Abbey. Cockington. Lanes about Marychurch. (E. B. t. 960.) P. VI.-VIII.

TRIBE II. POTENTILLIDÆ.

GEUM. AVENS.

G. urbanum (*common Avens.*)—Under hedges, by roadsides, banks, and borders of fields. Plant upright, with slightly branched stems, 1 to 2 feet high; root-leaves irregularly pinnate and lyrate, on long stalks; stem-leaves divided into 3 segments, all coarsely serrate. Flowers erect, small and yellow, petals spreading; calyx with 10 segments. Heads of fruit sessile, covered with silky hairs. Hedges by the side of the Ansti's Cove lane. Wood near Ansti's Cove. Cockington lanes, etc., frequent. (E. B. t. 1400.) P. VI.-VIII.

RUBUS. BRAMBLE, RASPBERRY.

1. **R. idæus** (*common Raspberry.*)—Woods and thickets; pretty generally distributed over Britain. Leaflets with close white down underneath. Flowers white, in long panicles at the ends of short branches. Fruit scarlet in a wild state. Road between Sandy Park and the Drewsteignton cromlech. Hedges in the lane at Gidleigh, near Chagford. Berry Pomeroy. (E. B. t. 2442.) Sh. VI.

2. **R. suberectus** (*upright Bramble.*)—Hedges, thickets, and boggy places. The *Rubi* have by some botanists been divided into a great number of distinct species, but it seems doubtful whether many of them are entitled to separate specific names. Bentham, in the preface to his 'Handbook,' says: "At any rate, if those minute distinctions by which the innumerable varieties of Brambles, of Roses, of Hawkweeds, or of Willows, have of late years been characterized, are really more constant and more important than the author's experience has led him to conclude, they cannot be understood without a more complete acquaintance with trifling, vague, and sometimes theoretical characters, than he himself has been able to attain, or than can ever be expected from the mere amateur." Most of the varieties of Bramble grow about and around Torquay, but this and the three following are all I have been able to distinguish. Forde Bog, near Newton. (E. B. t. 2572.) Sh. VII. VIII.

3. **R. fruticosus** (*common B.*)—In hedges, thickets, woods, and waste places, very abundant; varies considerably in the prickles and hairs, shape of the leaflets, and also in the colour of the flowers. Torquay, etc. (E. B. t. 715.) Sh. VII. VIII.

4. **R. córylifolius** (*Hazel-leaved B.*)—Thickets and hedges. Scarcely distinct from *R. rhamnifolius* and *R. carpinifolius.* Hedges about Torquay. (E. B. t. 827.) Sh. VI.–VIII.

5. **R. cæsius** (*Dewberry.*)—In open fields, thickets, hedge-banks, and borders of fields. Chudleigh. (E. B. t. 826.) Sh. VI.–IX.

FRAGARIA. STRAWBERRY.

F. vesca (*wood S.*)—In woods and bushy pastures, banks by roadsides, frequent. (A large variety, the *F. elatior* (E. B. t. 2197), is also frequently met with.) Flowers in Devonshire nearly the whole year. Very abundant in woods and lanes about Torquay and neighbourhood. (E. B. t. 1524, and S. t. 2742.) P. V.–VII.

POTENTILLA. CINQUEFOIL.

1. **P. anserina** (*Silver-weed.*)—Common by roadsides, in moist meadows, stony pastures, and waste places. Varies much in its degree of silkiness. Flowers large and bright yellow. Very abundant on roadsides and borders of fields about Torquay and Marychurch. (E. B. t. 861.) P. V.–VII.

2. **P. verna** (*spring C.*)—Pastures and waste places, most frequently in hilly and mountain districts. A small woody, procumbent plant, with flowers at the extremities of weak leafy branches. Torquay, etc. (E. B. t. 37.) P. IV.–VI.

3. **P. reptans** (*common creeping C.*)—In meadows, pastures, and roadsides. Common about Torquay and its vicinity. (E. B. t. 862.) P. VI.–IX.

4. **P. Tormentilla** (*Tormentil.*)—Moors and heathy places, frequent. Various places about Dartmoor. *Tormentilla officinalis*, Linn. (E. B. t. 863.) Var. β, *Tormentilla reptans*, Linn., grows on hedge-banks, borders of fields, and waste places about Torquay and Babbicombe. (E. B. t. 864.) P. VI. VIII.

5. **P. Fragariastrum** (*Strawberry-headed C.*)—On banks, dry pastures, and open woods, frequent. Banks by the sides of roads and lanes around Torquay and Marychurch. (E. B. t. 1785.) P. III.–V.

ALCHEMILLA. LADY'S-MANTLE.

A. arvensis (*field L. or Parsley Piert.*)—In meadows and pastures, and on wall-tops where there is any covering of soil. Flowers very small, green, and sessile, forming minute heads in the axils of the leaves, half-enclosed in the leafy stipules, frequent. Milber Down. About Torquay. (E. B. t. 597.) A. V.–VIII.

SANGUISORBA. BURNET.

S. officinalis (*great B.*)—In moist meadows and pastures, chiefly in a mountainous district. Stem from 1 foot to 2 feet high, branching upward; flowers much crowded, dark purple. Meadows near the Erme, at Ivybridge. Moor near Chagford. (E. B. t. 1312.) P. VI.–VIII.

POTERIUM. SALAD-BURNET.

P. Sanguisorba (*common S.*)—In dry pastures and clefts of limestone rocks. This plant generally puzzles young botanists by its curious compound-looking head of flowers. The lower flowers of the head are all male, the upper, female. Abundant on the Rock walk, Torquay. (E. B. t. 860.) P. VI.–VIII.

AGRIMONIA. AGRIMONY.

A. Eupatoria (*common A.*)—On roadsides, waste places, and borders of meadows, frequent. It varies considerably in the hairiness of its foliage, the size of its flowers, and the form of its ripe calyx. Warberry Hill. Hills above Meadfoot. Ansti's Cove Lane. Cockington. (E. B. t. 1335.) P. VI. VII.

A. ODORATA of Mill (Ansti's Cove, Mr. C. Parker): larger than the last, with scented flowers.

TRIBE IV. ROSEÆ.

Bentham, in his introductory remarks on the *Roses*, says that "even in the wild state, endeavours have been made to characterize so large a number of proposed species, that the confusion among them is almost as great as in the *Brambles*. The forms indigenous to Britain appear to be reducible to five types, which are probably real species. It must however be admitted, that the characters separating them are not so decided as could be wished, and that specimens will occasionally be found that the most experienced botanist will be at a loss to determine, and certainly not the less so if the number of British species be extended as proposed, to 15 or 20."

ROSA. ROSE, DOG-ROSE, SWEETBRIAR.

1. **R. spinosissima** (*Burnet-leaved R.*)—In dry bushy wastes, most common near the sea. Leaflets very small, from 7 to 9 to each leaf. Flowers small, whitish or pink, solitary, sweet-scented. Cliff walks near Ansti's Cove. Hope's Nose. (*Rosa pimpinellifolia*, Linn.) (E. B. t. 187.) Sh. V.

2. **R. rubiginosa** (*true Sweetbriar.*)—In hedges, thickets, and open, bushy places, in the south of England chiefly. More slender in its growth than the *Dog-Rose;* its prickles curved or hooked. The scent often very faint in the wild state. Ansti's Cove. Thickets at Chudleigh, where also *R. micrantha*, of Hooker and Arnott, is to be found. Lustleigh. (E. B. t. 991 and t. 2490.) Sh. VI. VII.

CALYCIFLORÆ. 43

3. **R. tomentosa** (*downy-leaved R.*)—Hedges and thickets, frequent. Probably a mere variety of *R. villosa* (E. B. t. 2459). Distinguished from the downy varieties of the Dog-Rose, chiefly by the globular fruit more or less covered with small fine prickles. The plant is also more erect and bushy. Both are found at Chudleigh. Chagford, near the gate leading to Gidleigh Park. (E. B. t. 990.) Sh. VI. VII.

4. **R. canina** (*common Dog-Rose.*)—Hedges and thickets, very common everywhere in England. Common about Torquay and neighbourhood. (E. B. t. 992.) *R. systyla*, given as a species by Hooker and Arnott, but considered by Bentham to be merely a variety of this, grows in thickets, etc., at Chudleigh. (*R. collina*, E. B. t. 1895.) Sh. VI. VII.

5. **R. arvensis** (*trailing Dog-Rose.*)—In hedges, woods, and thickets, very frequent. A much more trailing plant than *R. canina*, extending frequently to many feet. Flowers white, and without scent, 3 or 4 together. Fruit nearly globular and smooth. (E. B. t. 188.) Sh. VI. VII.

SUBORD. III. *POMEÆ.*

CRATÆGUS. HAWTHORN.

C. Oxyacantha (*Hawthorn, White-thorn, or May.*)—Woods, thickets, and hedges, abundant in England. Varies much in the form of the leaves, the down of its foliage and calyx, and also in the size and colour of its flowers. Common everywhere about the neighbourhood. (E. B. t. 2504.) T. V. VI.

PYRUS. PEAR, APPLE, SERVICE.

1. **P. Malus** (*Crab Apple.*)—Scattered over Britain, in woods and hedges. All the apples of our orchards belong to this species. Cliffs near Hope's Nose. Rocky valley near Torquay, etc. (E. B. t. 179.) T. V.

2. **P. torminalis** (*wild Service.*)—Woods and hedges, confined chiefly to the middle and south of England. A tall shrub, or moderately sized tree, with large white flowers in corymbs at the ends of short leafy branches. Near Chagford. Holne Chase. Ilsington. *Cratægus*, Linn. (E. B. t. 298.) T. IV. V.

3. **P. Aucuparia** (*mountain Ash or Rowan-tree.*)—Mountainous woods and hedges; distinguished from the two foregoing

by the regularly pinnate leaves. Sides of Tors, at Dartmoor. (E. B. t. 337.) T. V. VI.

4. **P. Aria** (*white Beam-tree*.)—Mountainous woods, particularly in limestone or chalky country. There are several varieties of this shrub. Park Hill, Torquay. (E. B. t. 1858.) T. V.

Ord. XXVIII. **ONOGRACEÆ.**

EPILOBIUM. WILLOW-HERB.

1. **E. hirsutum** (*great hairy W., Codlins-and-cream.*)—Sides of rivers, ditches, and in wet situations, frequent. Flowers large and handsome, rose-coloured. Whole plant downy. Torquay, road between Torre and Paignton. White variety grows at Paignton. (E. B. t. 838.) P. VII. VIII.

2. **E. parviflorum** (*small-flowered hairy W.*)—Localities same as the last. Very much like *E. hirsutum* on a small scale. Flowers very small. Road to Ansti's Cove. Meadows at back of Torre Abbey. Paignton. (E. B. t. 795.) P. VII. VIII.

3. **E. montanum** (*broad smooth-leaved W.*)—In waste and cultivated places, roadsides, woods, dry shady banks, walls and roofs of cottages, frequent. Stems from 6 inches to 1 foot high. Flower-buds erect or slightly nodding, flowers pink. Neighbourhood of Torquay and Marychurch. Milber Down. (E. B. t. 1177.) P. VI. VII.

4. **E. tetragonum** (*square-stalked W.*)—In damp places, and by sides of ditches. Stems from 1 to 2 feet high, with from 2 to 4 angles; leaves lanceolate, sessile; flower-buds erect; seeds oblong-obovate. (E. B. t. 1948.) This plant, as well as *E. obscurum*, Bab., which has much shorter capsules, grows by the side of ditches in the lane leading to Ansti's Cove. P. VII. VIII.

CIRCÆA. ENCHANTER'S NIGHTSHADE.

C. Lutetiana (*common E.*)—Woods and shady nooks, common. Whole plant graceful in the extreme; flowers white or pink, in elegant, slightly branched terminal racemes. Cockington and Shiphay lanes. Copse near Bishopstowe, etc. (E. B. t. 1056.) P. VI.–VIII.

CALYCIFLORÆ. 45

Ord. XXIX. HALORAGACEÆ.

HIPPURIS. MARE'S-TAIL.

H. vulgaris (*common M.*)—In shallow ponds and watery ditches. Whorls of about 8 leaves; the flowers are at the base of each of the upper leaves. Fruit a small, oblong, 1-seeded nut. Pools about Dartmoor. (E. B. t. 763.) P. vi. vii.

MYRIOPHYLLUM. WATER-MILFOIL.

M. spicatum (*spiked W.*)—Ditches and ponds. Leaves 4 in a whorl, and submerged. A slender spike protrudes from the water, bearing minute flowers arranged in whorls; the upper flowers are male, the lower female, very small. Fingle Bridge, near Chagford. (E. B. t. 83.) P. vi. vii.

Ord. XXX. LYTHRACEÆ.

LYTHRUM. PURPLE-LOOSESTRIFE.

L. Salicaria (*spiked P.*)—Wet ditches, watery and marshy places, common. Flowers reddish-purple, in large handsome spikes of dense whorls. Road to Ansti's Cove. Forde bog, near Newton, and generally diffused in moist situations. (E. B. t. 1061.) P. vii.–ix.

PEPLIS. WATER-PURSLANE.

P. Portula (*common W.*)—In wet ditches and watery places, not unfrequent. A small slightly branched annual, creeping and rooting at the base; leaves opposite; plant seldom more than 3 inches high. Flowers sessile in the axils of the leaves. Meadfoot, Torquay. Lustleigh. Buckland. (E. B. t. 1211.) A. vii. viii.

Ord. XXXI. TAMARICACEÆ.

TAMARIX. TAMARISK.

T. Anglica (*English T.*)—Rock, cliffs and shores by the sea.

A very graceful-looking plant, but most probably not indigenous. Banks on the sides of the road between Torquay and Paignton. (*T. Gallica,* Linn.) (E. B. t. 1318.) Sh. VII.

ORD. XXXII. CUCURBITACEÆ.

ORD. XXXIII. PORTULACEÆ.

MONTIA. BLINKS.

M. fontana (*water Blinks,* or *Chickweed.*)—By the edges of rills, and springy wet places, where the water does not become stagnant. A small, green, somewhat succulent plant, growing in dense tufts. Flowers solitary, or in small drooping racemes, white. Forde bog, near Newton. (E. B. t. 1206.) A. IV.-VIII.

ORD. XXXIV. PARONYCHIACEÆ.

ILLECEBRUM. KNOT-GRASS.

I. verticillatum (*whorled K.*)—In marshy and boggy grounds, confined to Devonshire and Cornwall. A minute plant, with spreading procumbent thread-like stems; leaves broadly ovate; flowers in axillary whorls, white. Boggy places in Dartmoor. (E. B. t. 895.) P. VII.

POLYCARPON. ALLSEED.

P. tetraphyllum (*four-leaved A.*)—Southern coast of England; particularly Devonshire and Dorsetshire. A much branched, spreading, prostrate plant. Flowers diminutive and numerous, in loose terminal cymes, growing in sandy situations near the sea. Kingswear. (E. B. t. 1031.) A. VI. VII.

SPERGULARIA. SANDWORT-SPURREY.

1. **S. rubra** (*field S.*)—Sandy and dry gravelly soils, common. Leaves narrow-linear; flowers extremely variable in size, generally

pink, sometimes nearly white. Paignton. Bovey. Sandy Park. (*Arenaria*, Linn.) (E. B. t. 852.) A. VI.–IX.

2. **S. marina** (*seaside S.*)—Common upon the seacoast. Larger in every respect than the last; but by most authors not considered to be specifically distinct. Cliffs by the bathing-cove, Torquay, and cliffs elsewhere around Torbay. (*Arenaria marina*, Sm.) (E. B. t. 958.) B. or P. VI.–VIII.

SPERGULA. SPURREY.

S. arvensis (*corn S.*)—In cultivated and waste places, too frequent in cornfields. Flowers small and white, on long slender stalks. Common over the whole county. (E. B. t. 1535.) A. VI.–VIII.

ORD. XXXV. CRASSULACEÆ.

COTYLEDON. PENNYWORT.

C. Umbilicus (*wall P.*)—On rocks, walls, and old buildings, abundant in western England. Remarkable for its shining, succulent, orbicular leaves, and erect stem, bearing a long raceme of pendulous yellowish-green flowers. Common on walls and rocks about Torquay, Ilsham, and Marychurch. (E. B. t. 325.) P. VI.–VIII.

SEMPERVIVUM. HOUSELEEK.

S. tectorum (*common H.*)—This is an introduced plant, and in England to be found only on old walls and cottage roofs. Leaves very thick and fleshy, the lower ones more than an inch in length, bordered by short stiff hairs, the upper leaves clothed with a viscid down. Flowers of a beautiful pink. Roofs of cottages at Cockington and Barton. (E. B. t. 1320.) P. VII.

SEDUM. ORPINE, STONECROP.

* 1. **S. Telephium** (*Live-long, or Orpine.*)—Borders of fields, hedge-banks, and shady copses. A very showy plant; stem from 1 foot to 2 feet high, with numerous oblong fleshy leaves, and bearing a leafy corymb of purple flowers; much cultivated

in gardens. Near Stover. Buckland. (E. B. t. 1319.) P. VII. VIII.

2. **S. Anglicum** (*English Stonecrop.*)—In rocky and stony places, usually not far from the sea. A small perennial, seldom exceeding 3 inches in height. Flowers white, but sometimes tinged with pink. Bovey Tracey. (E. B. t. 171.) A. VI.-VIII.

3. **S. album** (*white S.*)—Rocks, walls, and roofs of houses. Leaves a pale green; flowers white, or tinged with rose-colour, numerous and crowded. Meadfoot. Babbicombe. Marychurch, and other rocky places near Torquay. (E. B. t. 1578.) P. VII. VIII.

4. **S. acre** (*biting S.*, or *Wall-pepper.*)—On walls and rocks, in stony or sandy situations. Leaves small, thick, and ovoid, or nearly globular. Flowers of a bright shining yellow, in small terminal cymes. Rocks at the border of Daddyhole Plain. Road to Ansti's Cove. Babbicombe Down. Marychurch. (E. B. t. 839.) P. VI. VII.

Ord. XXXVI. GROSSULARIACEÆ.

RIBES. CURRANT, GOOSEBERRY.

1. **R. rubrum** (*common or red Currant.*)—In stony woods, frequent in Scotland and north of England, and occurring sometimes in southern England, whether wild or not is questionable. A branching shrub, from 3 to 4 feet high, without prickles; flowers small, greenish-white, in axillary racemes. Great wood on the Newton road, near the second gate leading into the lane opposite the cricket-ground. Berry Pomeroy wood. North Bovey. (E. B. t. 1289.) Sh. IV. V.

2. **R. Grossularia** (*common Gooseberry.*)—In thickets, open woods, and hedges; like the last, can scarcely be considered indigenous. A much branched, rather weak shrub, with numerous thorns beneath the small bunches of leaves. Flowers green; berry rather small and yellowish, with stiff hairs scattered over it. Shiphay lanes. Chagford, in a lane between Gidleigh and Gidleigh Common. Cliff walks at Ilsham. (E. B. t. 1292 and 2057.) Sh. IV. V.

Ord. XXXVII. SAXIFRAGACEÆ.

SAXIFRAGA. SAXIFRAGE.

S. tridactylites (*Rue-leaved S.*)—On walls and rocks, fre-

CALYCIFLORÆ. 49

quent in England. A pretty little plant from 2 to 5 inches high, simple or branched, and usually covered with a glandular down. Flowers small and white, growing on rather long flower-stalks. On old walls at Cockington and Paignton. Near Ilsham. Marychurch. (E. B. t. 501.) A. IV.-VII.

CHRYSOSPLENIUM. GOLDEN-SAXIFRAGE.

C. oppositifolium (*common G.*)—In moist shady places, by the sides of rivulets, common. Leaves all opposite. Flowers yellow, small and sessile, surrounded by small leaves. Cockington. Maidencombe. Coffin's Well. (E. B. t. 490.) P. IV. V.

Ord. XXXVIII. UMBELLIFERÆ.

In all the Umbellifers, a minute and careful examination of the flowers and seeds, more especially the latter, is absolutely necessary for the right discrimination of genera. In Hooker and Arnott's 'British Flora,' the student will find most accurate representations of the seeds of the different genera of this order, among the plates at the end of the volume (Tab. 1 to 3).

HYDROCOTYLE. WHITE-ROT.

H. vulgaris (*common W., or marsh Pennywort.*)—In bogs, marshes, edges of ponds and lakes, frequent. Flowers small and white. Fruits small, flat, and emarginate at the base. Forde bog, near Newton. Goodrington Marsh. Bogs in Dartmoor. (E. B. t. 751.) P. V.-VIII.

SANICULA. SANICLE.

S. Europæa (*wood Sanicle.*)—Woods and copses, frequent. Leaves for the most part radical. Flower-heads small and white. Fruit in small burrs. Wood near Bishopstowe. Before the building, plentiful in the Waldon Hill wood. (E. B. t. 98.) P. VI. VII.

ERYNGIUM. ERYNGO.

E. maritimum (*sea E., or Sea-Holly.*)—On sandy seacoasts, common. Whole plant stiff and rigid, glaucous or bluish.

E

Heads of flowers nearly globular, of a pale blue, and at first sight not having the appearance of an Umbellifer. The roots of this plant are sometimes candied, and are then pleasant to the taste, and have been esteemed stimulating and restorative. Paignton. Goodrington. (E. B. t. 718.) The *E. campestre*, which is very rare, grows at Devil's Point, Stonehouse, near the Victualling Office, but is now almost extinct. (E. B. t. 57.) P. VII. VIII.

APIUM. CELERY.

A. graveolens (*Smallage, or wild C.*)—Marshes and ditches, especially near the sea. Stem from 1 to 2 feet high, furrowed. Flowers small and white; fruit very small, roundish-ovate. The umbels often sessile, when peduncled bearing but few flowers. Sides of ditches in meadows behind Torre Abbey. Kingskerswell, near the church. Banks of Dart, near Totness. (E. B. t. 1210.) P. VI.–VIII.

PETROSELINUM. PARSLEY.

P. segetum (*corn P.*)—In fields and wastes places in central and southern England, frequent. A much branched slender plant, from 9 inches to 2 feet high; leaves chiefly radical, pinnate, with from 5 to 10 pairs of sessile, cut, and serrate segments. Umbels irregular, partial umbels containing very few small and white flowers. Fruit strongly ribbed. Fields about Torquay and Marychurch. Chudleigh, etc. *Sison*, Linn. (E. B. t. 228.) B. VI.–VIII.

TRINIA. HONEWORT.

T. vulgaris (*common H.*)—In dry, arid, limestone wastes, rare, confined to the south-western counties in England. Stock short and thick, almost woody, forming a long tap-root at its base; stems erect, stiff and angular, with many spreading branches, nearly a foot high. Leaves cut into nearly linear segments. Flowers small and numerous, white. Fruit obtuse. Berry Head. *Pimpinella dioica*, Sm. (E. B. t. 1209.) P. V. VI.

HELOSCIADIUM. MARSH-WORT.

H. nodiflorum (*procumbent M.*)—In marshes, meadows,

wet ditches, and sides of rivulets. Abundant in England. Stems creeping and rooting at the base. Umbels nearly sessile. Fruit broadly ovate. Wet meadows near Torre Abbey. *Sium*, Linn. (E. B. t. 639.) P. VII. VIII.

2. **H. inundatum** (*least M.*)—In swamps, shallow ponds, and pools that are dried up in summer. Much like the last, but smaller. Umbels usually of two rays. Fruit large in proportion to the size of the plant. Paignton marsh. Marychurch. Bovey Heath. (*Sison*, E. B. t. 227.) P. VI. VII.

SISON. BASTARD STONE-PARSLEY.

S. amomum (*hedge bastard S.*)—Under hedges and in thickets, frequent. Stem about 2 feet high. Leaves pinnate below, upper ones cut into narrow segments. Umbels on slender footstalks, with but few white flowers on short pedicels. Fruit roundish-ovate. Common everywhere. (E. B. t. 954.) P. VIII.

ÆGOPODIUM. GOUT-WEED.

Æ. Podagraria (*common G., or Bishop's-weed.*)—In moist woods and copses. An upright plant, stem about 18 inches high. Root-leaves once, upper ones twice ternate. Umbels large, with numerous white flowers. Fruit oblong, about one-sixth of an inch in length. North Bovey. About Totness. Moreton, *Fl. D.* (E. B. t. 940.) P. VI. VII.

BUNIUM. EARTH-NUT.

B. flexuosum (*common E.*)—In woods and pastures, common. The tuber is sweet and wholesome, and is much relished by children and pigs. The plant slender, bearing very narrow leaves with linear segments; umbels terminal. Fruit oblong, slightly ribbed. Woods, thickets, and waste places about Torquay, etc. (E. B. t. 988.) P. V. VI.

PIMPINELLA. BURNET-SAXIFRAGE.

P. saxifraga (*common B.*)—In pastures, on banks and road-sides, frequent. A plant from 1 foot to 2 feet high, with very variable leaves. Umbels terminal, bearing white flowers. Fruit ovate, surmounted with the swollen base of the reflexed styles, glabrous. Torquay. Marychurch, etc. (E. B. t. 407.) P. VII. VIII.

BUPLEURUM. HARE'S-EAR.

B. aristatum (*narrow-leaved H.*)—In England only in the neighbourhood of Torquay. A very small plant, from 1 to 2 or 3 inches high, slightly branched; leaves narrow-linear, 3-nerved. Umbels very small, terminal, much involucrated, bearing minute yellow flowers. Fruit ovate-oblong. Mound by Daddyhole Plain, Torquay, overlooking the quarry. Sometimes on Park Hill, near the stone seat. Ilsham and Babbicombe Downs. *B. Odontites,* Linn. (E. B. t. 2468.) A. VI. VII.

ŒNANTHE. WATER-DROPWORT.

1. **Œ. fistulosa** (*common W.*)—In wet meadows, ditches, and rivulets, common. Plant from 2 to 3 feet high or more; all the branches as well as the stem remarkably tubular. Leaves twice pinnate below; the leaves of the stalks bear only a few pinnate segments, with linear lobes. Umbels terminal, the centre one having only 3 rays. Fruit in compact globular heads, narrowed at the base, and crowned by the narrow teeth of the calyx and the longer rigid styles. Exminster marsh. Chudleigh. Powderham marshes. (E. B. t. 363.) P. VII.-IX.

2. **Œ. pimpinelloides** (*callous-fruited W.*)—Meadows and pastures. Stem erect and firm; roots fibrous, swelling into ovoid tubers. Leaves much divided. Umbels many-rayed, with the smaller umbels much crowded. Flowers sometimes having a faint yellowish-green colour. Fruit an ovoid cylinder crowned with long styles. Paignton (Mr. Earle). Exminster. Near Exmouth, *Fl. D.* (E. B. t. 347.) P. VI. VII.

3. **Œ. crocata** (*Hemlock W.*)—In wet ditches and by riversides, frequent. A strong and branched species, from 3 to 5 feet high. Leaves 2 or 3 times pinnate. Umbels on long terminal peduncles; flowers white. Fruit somewhat corky; the ribs broad but not prominent. Torquay and Marychurch. Chudleigh. Moreton. (E. B. t. 2313.) P. VII.

ÆTHUSA. FOOL'S-PARSLEY.

Æ. Cynapium (*common F.*)—Abundant in England, in fields and under hedges. An upright, very leafy plant; the lower leaves thrice-pinnate, the segments more or less cut into narrow lobes. Umbels on long peduncles. Flowers white. Plant possesses a nauseous smell. Common in the neighbourhood. (E. B. t. 1192.) A. VII.–VIII.

SIUM. WATER-PARSNIP.

S. angustiflorum (*narrow-leaved W.*)—In wet ditches and shallow streams, common. Not so tall a plant as *S. latifolium*, but more branched and leafy; seldom attains to 2 feet in height. Stem erect; leaflets unequally lobed and serrate. Umbels numerous, small, on short footstalks. Flowers white. Fruit broadly ovoid, slightly compressed laterally. Marshy meadow near Torre Abbey, not far from the high-road. Meadow behind Forde House, Newton. Banks of the Dart, near Totness, *Fl. D.* (E. B. t. 139.) P. VIII.

FŒNICULUM. FENNEL.

F. vulgare (*common Fennel.*)—On dry rocky banks, near the sea. Stem erect, much-branched; leaves three or four times pinnate, with very narrow linear segments. Umbels large, with many rays; flowers deep-yellow; fruit oblong. The whole plant has a very strong, but not unpleasant smell. Cliffs above Meadfoot, and banks by the side of the Paignton road. (E. B. t. 1208.) P. VII. VIII.

SILAUS. PEPPER-SAXIFRAGE.

S. pratensis (*meadow P.*)—In meadows and moist pastures, by waysides, beneath damp hedges. Stem erect, from 1 to 3 feet high, slightly branched; leaves thrice pinnate; leaflets linear-lanceolate. Flowers pale yellow. Fruit oval. Whole plant fetid when bruised. Babbicombe road, beneath the wall opposite the the Ansti's Cove lane. Barton. (E. B. t. 2142.) P. VI.–IX.

CRITHMUM. SAMPHIRE.

C. maritimum (*sea Samphire.*)—In clefts of rocks near the sea. Rarely more than a foot high. The young branches, leaves, and umbels thick and fleshy; leaves two or three times divided. Umbels many-rayed; flowers yellow and small; fruit elliptical, about 3 lines long. Rocks by the sea at Babbicombe, and around Torbay generally. Teignmouth. Dawlish. (E. B. t. 819.) P. VIII.

ANGELICA. ANGELICA.

A. sylvestris (*wild A.*)—In moist woods and marshy places, particularly near streams. Abundant in England. A tall, strong, branching plant, 3 or 4 feet high, with thick stems. Lower leaves large, doubly pinnate, with segments more than 2 inches long. Umbels large, the main ones often with 30 or 40 rays. Flowers pinkish-white. Fruit flat, with 2 wings on each side. Ansti's Cove lane, near the stile leading to Kent's Cavern. (E. B. t. 2561.) P. VII.-IX.

HERACLEUM. COW-PARSNIP.

H. Sphondylium (*common C., or Hog-weed.*)—In meadows, pastures, hedges, and thickets, common. A tall coarse plant, more or less covered with short stiff hairs. Leaves pinnate, with several large and broad segments. Umbels large, with about 20 rays. Flowers white, sometimes reddish. Fruit flat, with a broad border. Hedges and pastures, very common. Grows most abundantly on the Rock walk, Torquay. (E. B. t. 939.) P. VII.

CONIUM. HEMLOCK.

C. maculatum (*common H.*)—On the banks of streams, waste places, under hedges and walls, borders of fields, not unfrequent over Britain. An upright, branching plant, from 3 to 6 feet high, giving out a nauseous smell when bruised; recognized immediately by its polished stem spotted with purple. Leaves large, and divided into many deeply cut lanceolate segments. The larger and smaller umbels many-rayed. Flowers white. Fruit broadly-ovate, with waved ridges. Whole plant highly poisonous. Hedges and waste places about Torquay and Marychurch, etc. Torre Abbey meadow. (E. B. t. 1191.) B. VI. VII.

SMYRNIUM. ALEXANDERS.

S. Olusatrum (*common A.*)—In meadows and waste places, waysides and among ruins, especially near the sea. Plant with a stout, furrowed stem, from 2 to 4 feet high; lower leaves two or three times, upper ones once ternate, of a yellowish-green colour, having a broad membranous base. Umbels large, roundish, and many-rayed, on stout footstalks; flowers small, greenish-yellow. Fruit of 2 nearly round lobes, with prominent ribs, nearly black when ripe. Rock walk. Marychurch road. Paignton road, etc. (E. B. t. 230.) A. or B. v. vi.

SCANDIX. SHEPHERD'S-NEEDLE.

S. Pecten (*common S., or Venus's-comb.*)—In waste places and cornfields, abundant. Plant branching or spreading, generally hairy, from 6 inches to a foot high. Leaves 2 or 3 times pinnate. Umbels irregular, but usually 2- or 3-rayed. Flowers white. Fruit laterally compressed, with a long beak, 3 or 4 times as long as the rough fruit, having much the appearance of the tooth of a comb. *S. Pecten-Veneris*, Linn. (E. B. t. 1397.) A. v.-ix.

ANTHRISCUS. BEAKED-PARSLEY.

A. sylvestris (*wild B.*)—One of the commonest Umbellifers in England. Under hedges and along the borders of fields. Stem hairy below, smooth above, slightly swollen beneath the joinings; leaves twice pinnate. Umbels at first rather drooping, numerous, with 8 or 10 rays. Flowers small and white. Fruit smooth and shining, contracted at the top, with no apparent beak. Frequent; Torquay, Marychurch, Chudleigh, etc. *Chærophyllum*, Linn. (E. B. t. 752.) P. iv.-vi.

CHÆROPHYLLUM. CHERVIL.

C. temulentum (*rough C.*)—Hedges and thickets, common. Plant from 2 to 3 feet high, rough with short hairs; leaves 2 or 3 times pinnate, with numerous pinnatifid segments, more or less hairy. Umbels at first drooping, of many rays, covered with bristly hairs. Flowers white. Fruit narrow, flask-shaped, ribbed. Fields and hedges, common. Rock walk. Marychurch road, etc. (E. B. t. 1521.) P. vi. vii.

DAUCUS. CARROT.

1. **D. Carota** (*wild C.*)—Pastures, waste places, and borders of fields, common. Plant upright, with a fusiform root; stem from 9 inches to 2 feet high. Leaves 3-pinnate, with numerous linear acute segments. Umbels large and compact, of numerous white flowers, with usually a central flower of deep crimson. Fruit oblong and covered with prickles. Daddyhole Plain. Warberry Hill. Marychurch, etc. (E. B. t. 1174.) B. VI.–VIII.

2. **D. maritimus** (*seaside C.*)—Smaller than the last, with the leaves somewhat fleshy, with shorter segments, footstalks thicker. Umbels convex or flat when in seed; flowers entire, white, or with a slight tinge of red. Prickles of the fruit shorter. Seacoasts in the south of England. Cliffs at Meadfoot and Daddyhole Plain, and along the coast. *D. gummifer*, Bab. (E. B. t. 2560.) B. VII. VIII.

TORILIS. HEDGE-PARSLEY.

1. **T. Anthriscus** (*upright H.*)—In hedges, by roadsides and waste places. Stem erect, 2 or 3 feet high. Leaves once, or the lower ones twice pinnate. Umbels on long stalks; flowers reddish or white. Fruit a small burr, rough with incurved but not hooked bristles. Waldon Hill. Hope's Nose, and common in hedges, etc. (*Caucalis*, E. B. t. 987.) A. VII. VIII.

2. **T. infesta** (*spreading H.*)—Fields, banks, and roadsides, common. Like the last, but smaller and more spreading. Flowers yellowish-white. Fruit with straight bristles, with a small hook at the top. Cultivated fields near Torquay, etc. (*Caucalis*, E. B. t. 1314.) A. VII. VIII.

3. **T. nodosa** (*knotted H.*)—Waste places and by roadsid especially in dry gravelly soil. Stem prostrate. Leaves twice pinnate. Umbels forming little heads. Flowers reddish. Fruit smaller than in the two preceding, the outer ones having bristles but the inner tubercles only. Meadfoot Cliffs. Marychurch. (*Caucalis*, E. B. t. 199.) A. V.–VII.

ORD. XXXIX. **ARALIACEÆ.**

ADOXA. MOSCHATEL.

A. moschatellina (*tuberous M.*)—Woods and shady hedge-

CALYCIFLORÆ. 57

banks, common in Britain. A small plant, of a delicate-green colour in all its parts; stem about 8 inches high; leaves coming from the root on long footstalks, two or three times divided, with deep 3-lobed segments. Flower-stalk also radical, 4 or 6 inches high, bearing curious-looking little green flowers, in a globular head of five, four of which are placed laterally and are crowned with the fifth. Berry green and fleshy, containing usually but one seed. Bank near the pools by Lisburn Crescent. Totness. Chudleigh. (E. B. t. 453.) P. IV. V.

HEDERA. IVY.

H. Helix (*common Ivy.*)—In woods, round the stems of trees, on rocks and old walls. A very common, woody, evergreen climber, with leathery oval or heart-shaped leaves of a shining dark green, often streaked with white veins. Flowers small, of a greenish-yellow. Berries smooth and black. Woods, walls, and rocks about Torquay and Babbicombe, etc. (E. B. t. 1267.) Sh. X. XI.

ORD. XL. **CORNACEÆ.**

CORNUS. CORNEL, DOGWOOD.

C. sanguinea (*wild C. or D.*)—In hedges and thickets, abundant in the south of England. An arborescent shrub, from 5 to 7 feet high, with smooth red branches; leaves a broad oval shape, opposite and stalked, becoming a deep red before they drop. Flowers in cymes, numerous, of a dull white; berries dark purple, extremely bitter. Woods and hedges, frequent. Ansti's Cove. Marychurch. (E. B. t. 249.) Sh. VI.

ORD. XLI. **LORANTHACEÆ.**

VISCUM. MISTLETOE.

V. album (*common M.*)—Common in southern and western England. Parasitic, generally on Apple-trees, but sometimes on the Oak. Stems woody when old, with repeatedly-forked branches; leaves obtusely lanceolate, like the branches, of a sickly-green colour. Flowers yellowish-white, nearly sessile in the forks of the branches. Berries white, semi-transparent. On Apple-trees in the orchard at Ilsham. (E. B. t. 1470.) P. III.-V.

58 CALYCIFLORÆ.

Ord. XLII. CAPRIFOLIACEÆ.

SAMBUCUS. ELDER.

1. **S. Ebulus** (*dwarf E.*, *or Danewort.*)—Roadsides, wastes, and stony places. A small shrub, from 2 to 3 feet high, with a rough, angular, and furrowed stem. Leaves pinnate, with serrated leaflets; cymes with three principal branches. Flowers purplish. Berries round and nearly black. The plant disagreeable to the smell. Marychurch. (E. B. t. 475.) P. VII. VIII.

2. **S. nigra** (*common E.*)—In woods, thickets, and waste places, frequent. A small tree, with the stem and branches hollow and filled with white pith; leaves pinnate, with ovate serrated segments. Cymes with 5 main branches. Flowers white or cream-coloured, strongly scented. Fruit black. Hedges and woods about Torquay and Marychurch. (E. B. t. 476.) T. VI.

VIBURNUM. GUELDER-ROSE.

1. **V. Lantana** (*mealy G.*, *or Wayfaring-tree.*)—Woods and hedges, especially in a limestone soil, frequent. A large and much branched shrub, with the young shoots and leaves very downy. Leaves very broad and serrate, veined, with their under sides mealy. It bears large dense cymes of white flowers, which ripen into purplish-black berries. Ansti's Cove. Meadfoot Cliffs, Torquay. Totness. Chudleigh. (E. B. t. 331.) Sh. IV. V.

2. **V. Opulus** (*common G.*)—In hedges and coppices, frequent. A small tree, with opposite branches; leaves 3- to 5-lobed. Flower-cymes large, with white flowers, the outer ones radiant and barren, the inner ones fertile. Berries red. Bradley woods, near Newton. Forde bog. Chudleigh. Holy Street, near Chagford, in a bushy place by the Teign. (E. B. t. 332.) T. VI. VII.

LONICERA. HONEYSUCKLE.

L. Periclymenum (*common H.*, *or Woodbine.*)—Woods and hedges, very frequent. A woody climbing plant, spreading itself over bushes, trees, and rocks to a great extent. Leaves broadly ovate, the lower ones stalked, the upper ones closely sessile. Flowers pale yellow or reddish, in sessile, terminal heads. Berries red. Park Hill wood. Bushy places about Meadfoot. Ansti's Cove, etc. (E. B. t. 800.) Sh. VI.-IX.

Ord. XLIII. RUBIACEÆ.

RUBIA. MADDER.

R. peregrina (*wild M.*)—In dry woods and rocky places in the south-west of England. A very straggling plant, frequently trailing over hedges to the extent of many feet, fastening itself by its recurved prickles at the edges of its whorled leaves. Flowers very small and green, in loose panicles. Fruit a small two-lobed rough berry. In nearly every hedge about Torquay. Plentiful on the Rock walk. (E. B. t. 851.) P. VI.-VIII.

GALIUM. BEDSTRAW.

1. **G. verum** (*yellow B.*)—On dry banks and pastures, sandy places and seashores, common. Stem from 6 inches to 1 foot high; leaves in whorls of from 6 to 8, slightly rough on their edges. Flowers small and numerous, forming a yellow terminal panicle. Torquay and Paignton, very common. (E. B. t. 660.) P. VII. VIII.

2. **G. cruciatum** (*Crosswort B.*, *Mugwort.*)—Hedges and thickets, frequent. Leaves 4 in a whorl, hairy. Stem simple, from 1 to 2 feet high. Flowers small and yellow, in little leafy clusters. Fruit small and smooth. Warberry Hill. Barton. Teignmouth road, etc. (E. B. t. 143.) P. V. VI.

3. **G. saxatile** (*smooth heath B.*)—In open heaths and mountainous pastures, very frequent. Plant 5 or 6 inches high, much branched, and leafy, with about 6 leaves in a whorl, the lower leaves small and ovate, the upper narrow, all having a little point at their extremity. Flowers in terminal panicles, numerous. Fruit small and granulated. Babbicombe Down. Forde, near Newton. Milber Down. Open places about Chagford, etc. (E. B. t. 815.) P. VII. VIII.

4. **G. uliginosum** (*rough marsh B.*)—Swampy meadows and sides of ditches. Distinguished from *G. saxatile* by its stem being rough on the angles. Terminal panicles small and few-flowered. Fruit dark-brown. Berry Pomeroy woods. Banks of Dart, near Totness, *Fl. D.* Banks of Teign, near Gidleigh. (E. B. t. 1972.) P. VII. VIII.

5. **G. palustre** (*white water B.*)—In marshes and wet places, frequently growing completely in the water. A weak and slender plant; stems about 1 foot long, with usually leaves of 4 in a whorl, without any points at their tips. Flowers small and white.

Fruit small. Frequent in wet places. Forde bog, near Newton. (E. B. t. 1857.) P. VII. VIII.

6. **G. Aparine** (*Goose-grass or Cleavers.*)—In hedges and thickets, etc., very abundant. A straggling, scrambling plant, often several feet long, clinging to the branches of bushes by the rough prickles on the angles of its stem and the edges of its leaves. Leaves from 6 to 8 in a whorl. Flowers greenish-white, in cymes. Fruit covered with little hooked bristles, fastening readily to the coats of animals. Common in nearly every hedge. (E. B. t. 816.) A. VI.-VIII.

7. **G. Mollugo** (*great hedge B.*)—In hedges, thickets, and pastures. Stems from 1 to 3 feet long. Leaves 8 in a whorl, rough, terminated by a little point. Flowers in panicles, white and numerous. Fruit small and smooth. Woods and hedges, very common. (E. B. t. 1673.) P. VII. VIII.

SHERARDIA. SHERARDIA, FIELD MADDER.

S. arvensis (*blue S.*)—In cornfields and dry pastures. A small plant, seldom so much as 6 inches high. Leaves about 6 in a whorl, linear, rough on the edges, and ending in a fine point. Flowers small, in terminal heads, either blue or pink. Has much the appearance of a small *Galium*. Fields about Torquay and Marychurch, very common. (E. B. t. 891.) A. V.-VII.

ASPERULA. WOODRUFF.

1. **A. odorata** (*sweet W.*)—In woods and shady places. Plant from 6 inches to a foot high. Leaves generally 8 in a whorl, lanceolate, slightly rough at the edges. Flowers white, in a loose cyme. The whole plant, especially when drying, smells like new-made hay. Chudleigh. Lustleigh. Exmouth. *Fl. D.* (E. B. t. 755.) P. V. VI.

2. **A. cynanchica** (*small W., Sqinancy-wort.*)—On dry, warm banks, pastures, and limestone hills. A smooth plant, with but few leaves, 6 or 8 inches high. Leaves narrow, linear, the lower ones 4 and the upper 2 in a whorl. Flowers white, with a tinge of lilac. Fruit small. Warberry Hill. Berry Head. (E. B. t. 33.) P. VI. VII.

Ord. XLIV. VALERIANACEÆ.

CENTRANTHUS. SPUR-VALERIAN.

C. ruber (*red Spur-Valerian.*)—On old walls, rocks, and in stony places. Plant from 12 to 18 inches or 2 feet high; stems numerous; leaves smooth, ovate-lanceolate, the lower ones entire, upper often toothed. Flowers red, sometimes white, in dense cymes. Tube of the corolla with a spur. Fruit with a feathery pappus. Common on rocks, and on nearly every old wall about Torquay. Exeter, Dawlish, Teignmouth, Dartmouth, *Fl. D. Valeriana rubra*, Linn. (E. B. t. 1531.) P. VI.-IX.

VALERIANA. VALERIAN.

1. **V. dioica** (*small marsh V.*)—Marshy meadows, not unfrequent. Several erect flowering stems, 6 to 8 inches high, are sent off from the same root. The root-leaves on long stalks, ovate, entire; those of the stem pinnate. Flowers in terminal corymbs, small, pale-rose colour. Milber Down. Forde bog, near Newton. Marychurch, *Rev. A. Neck, in Fl. D.* (E. B. t. 628.) P. V. VI.

2. **V. officinalis** (*great wild V.*)—Sides of ditches and rivers, in moist situations and damp woods, very frequent. Stems much larger than the last, from 2 to 4 feet high. Leaves pinnate, with from 9 to 21 lanceolate segments; the upper leaves few and distant. Flowers small, pinkish-white. A very handsome-looking plant. Side of a stream at the back of Forde House, Newton. Forde bog. Side of Stover canal. (E. B. t. 698.) P. VI. VII.

FEDIA. CORN-SALAD.

1. **F. olitoria** (*common C.*, or *Lamb's-Lettuce.*)—Waste places, hedges, and cornfields. An insignificant-looking plant, seldom more than 6 inches high, branching at the base and repeatedly forked. Leaves opposite, oblong. Flowers in terminal heads, of a pale-blue colour. Fruit roundish, 3-celled. Common in fields on the Warberry Hill, etc. (E. B. t. 811.) *Valerianella*, Bab. A. V. VI.

2. **F. dentata** (*smooth narrow-fruited C.*)—Hedge-banks and cornfields, not so common as the last. Stem erect; leaves small and narrow. Flowers flesh-coloured, single, in the division of

the stem. Fruit oval, crowned with the 4-toothed calyx. *Valerianella*, Bab. Warberry Hill. Marychurch. Moreton, N. Bovey, Lustleigh, *Fl. D.* (E. B. t. 1370.) A. VI. VII.

Ord. XLV. DIPSACEÆ.

DIPSACUS. TEASEL.

1. **D. fullonum** (*fuller's T.*)—On hedge-banks and waste places, *scarcely wild*, differing only from the next in the scales of the seed-vessel being hooked at the extremity. Used in the dressing of cloth. Hedge between Exminster and Alphington, *Fl. D.* (E. B. t. 2080.) B. VIII. IX.

2. **D. sylvestris** (*wild T.*)—By roadsides and in waste places, frequent. Plant 4 or 5 feet high; the stems, midribs of the leaves, leafstalks, and involucres set with numerous prickles; leaves opposite, sessile, coarsely toothed. Flowers pale lilac; the flower-heads large and conical. Scales of the seed-vessel straight, ending in a fine point. Between Meadfoot and Hope's Nose. Paignton. Chudleigh. Exminster, Exeter, *Fl. D.* (E. B. t. 1032.) B. VIII. IX.

SCABIOSA. SCABIOUS.

1. **S. succisa** (*devil's-bit S.*)—Meadows and pastures. Plant from 12 inches to 2 feet high, with stalked, oblong, entire radical leaves, and with from 1 to 5 heads of deep-blue flowers, on long footstalks. Corolla 4-cleft. Root abruptly broken off, as if it had been bitten, whence its English name. Ansti's Cove, on the rocks dividing the white beach from the cove. (E. B. t. 878.) P. VII.-X.

2. **S. Columbaria** (*small S.*)—Waste places and pastures, very frequent. Plant 1 to 2 feet high. Leaves pinnate; stem-leaves few. Flowers pale purplish-blue, 5-lobed. Common in fields and waysides about Torquay and Marychurch. Teignmouth. Dawlish, etc. (E. B. t. 1311.) P. VII. VIII.

KNAUTIA. KNAUTIA.

K. arvensis (*field K.*)—Fields and pastures, common. Stem from 2 to 3 feet high, hairy and branched, with but few pinnatifid leaves, but with many lanceolate radical leaves. Flowers purple, in large convex heads, on long stalks; corolla 4-cleft. In cornfields and pastures everywhere. (E. B. t. 659.) *Scabiosa*, Linn. P. VII.-IX.

ORD. XLVI. COMPOSITÆ.

The student will, in investigating the plants belonging to this Order, as in the case of the Umbellifers, be greatly assisted by a reference to the plates in Hooker and Arnott's British Flora, t. 3 A, and t. 4 and 5.

TRIBE I. CICHORACEÆ.

TRAGOPOGON. GOAT'S-BEARD.

T. pratensis (*yellow G.*)—Meadows and pastures, frequent. A coarse grassy-looking plant, from 1 to 2 feet high, with long, undivided, channelled leaves; flower-stalks slightly thickened at the summits, with a single head of yellow flowers. Head of seeds large; pappus very feathery, raised upon a long stalk. The flower closes every day before noon. Turf in Apsley House ground. Park Hill, just by the flagstaff. Meadow in front of Torre Abbey. Chudleigh. Totness, etc. (E. B. 434.) B. or P. VI. VII.

HELMINTHIA. OX-TONGUE.

H. echioides (*bristly O.*)—On hedge-banks, borders of fields and waste places. A coarse plant, growing to 2 or 3 feet high, with a rough, hairy stem, and large, lanceolate, clasping leaves, very rough on their upper sides. Flower-heads small, and crowded on short stalks; flowers yellow. Seeds with a feathery pappus. Walks above Meadfoot. Cliffs by the side of the New road, Torquay. Babbicombe. Road between Totness and Stoke Gabriel. Cliffs at Exmouth, *Fl. D. Picris*, Linn. (E. B. t. 972.) A. VII-IX.

PICRIS. PICRIS.

P. hieracioides (*Hawkweed P.*)—Under hedges, by roadsides and waste places, common. Plant from 2 to 3 feet high; stem rough with short minutely-hooked hairs; leaves lanceolate and coarsely toothed. The lower leaves with stalks; the upper clasping the stem. Flowers yellow; the heads collected into an imperfect corymb. Meadfoot Cliffs and Lincombe Hill. Roadsides about Torquay. (E. B. t. 196.) B. VII.-IX.

APARGIA. HAWKBIT.

1. **A. hispida** (*rough H.*)—In meadows and pastures, common. The whole plant more or less hairy. Leaves growing from the root, long and narrow and coarsely toothed. Flower-stalk erect, about 6 inches high, swollen at the top, with one large flower-head. Flowers yellow. Involucre of flowers hairy. Frequent about Torquay. Teignbridge. Sands at Exmouth, *Mr. Jacob, in Fl. D.* (E. B. t. 554.) P. VI.-IX.

2. **A. autumnalis** (*autumnal H.*)—Fields, roadsides and waste places, very frequent. Leaves all radical, linear-lanceolate, deeply toothed, not hairy. Flower-stems erect, with one or two single-headed branches, with small scales. Flowers yellow, those of the circumference inclining to red. Near Barton. (E. B. t. 830.) P. VIII.

THRINCIA. THRINCIA.

T. hirta (*hairy T.*)—In dry open pastures and moors, frequent. Leaves coming from the root, lance-shaped, sometimes quite entire, at others slightly toothed, somewhat hairy; flower-stalks about 6 inches high, each with a single head of bright yellow flowers. Root having the appearance of being bitten off. Neighbourhood of Torquay, etc. *Hedypnois.* (E. B. t. 555.) P. VII.-IX.

HYPOCHŒRIS. CAT'S-EAR.

H. radicata (*long-rooted C.*)—Meadows, pastures, and waysides, frequent. Leaves radical, hairy, spreading, long and narrow, more or less indented. Flower-stem from 1 to 2 feet high, sometimes branched, thickening upwards, and bearing a large head of yellow flowers. Fruits beaked. Common everywhere in the neighbourhood. (E. B. t. 831.) P. VII.

SONCHUS. SOW-THISTLE.

1. **S. palustris** (*tall marsh S.*)—In marshes, the edges of

ponds and wet ditches. A tall plant, from 5 to 8 feet high, with long, narrow, clasping leaves, 8 or 10 inches long, with acutely pointed auricles. Flower-heads corymbose; flowers yellow and numerous. Banks of the Exe, near Powderham, *Fl. D.* (E. B. t. 935.) P. VII. VIII.

2. **S. arvensis** (*corn S.*)—Frequent in cornfields. Plant from 2 to 3 feet high. Leaves toothed and clasping the stem, with short and blunt auricles; the lower leaves stalked. Flower-heads in loose terminal panicles; flowers bright-yellow. The whole plant rough with brown or black glandular hairs. Feathers of the seed white and silky. Warberry Hill, Torquay, etc. (E. B. t. 674.) P. VIII. IX.

3. **S. oleraceus** (*common annual S.*)—Cultivated ground and waste places, very common. Plant from 1 to 3 or 4 feet high, with a thick hollow stem, generally smooth; lower leaves stalked and deeply divided; upper ones lanceolate and clasping the stem. Flower-heads rather small, set in an umbel-like arrangement; flowers pale-yellow; seeds bearing a snow-white pappus. Fields and banks about Torquay and Marychurch. (E. B. t. 843.) A. VI.–VIII.

4. **S. asper** (*sharp-fringed annual S.*)—Growing in similar situations to the last, of which it appears to be merely a variety, having the leaves more sharply toothed, and the auricles broader, and set with more prickly teeth. Fruit without the transverse wrinkles of *S. oleraceus.* Chudleigh. (E. B. S. t. 2765, 2766.) A. VI.–VIII.

CREPIS. HAWK'S-BEARD.

C. virens (*smooth H.*)—In pastures, on dry banks, roofs of cottages, and by roadsides, common. Plant from 1 to 3 feet high, branched and nearly smooth. Leaves somewhat lanceolate, pinnatifid. Root-leaves stalked, upper ones linear and clasping. Flower-heads small, in loose panicles; flowers yellow. Seeds oblong, shorter than their pappus. Common about Torquay, etc. (*C. tectorum*, E. B. t. 1111.) A. VI.–IX.

LEONTODON. DANDELION.

L. Taraxacum (*common D.*)—Waysides, meadows, and pastures, common. A plant of very variable appearance; leaves all growing from the root, generally deeply pinnatifid, with broad triangular lobes, pointing backwards; but sometimes long and lance-shaped and scarcely, if at all, cut. The flower-stalk proceeds also from the root, and rises from 2 to 6 or 8 inches high, bearing a large head of yellow florets, which when ripe forms a globular head of delicate feathered seeds. This and var. β grow abundantly in the neighbourhood. (E. B. t. 510, 553.) P. III.-X.

HIERACIUM. HAWKWEED.

1. **H. Pilosella** (*common mouse-ear H.*)—In dry pastures, on banks and roadsides. Plant from 3 to 6 inches high. Root giving off creeping scions; leaves radical, lanceolate, tapering towards the stem, having a few hairs on their upper surface, but with a white down on their under sides. Flower-stalks with a single head of light-yellow flowers, frequently tinged with red on the outside. Daddyhole Plain. Babbicombe, etc. (E. B. t. 1093.) P. V.-VIII.

2. **H. murorum** (*wall H.*)—On banks and old walls, in meadows and pastures. Flower-stems erect, from 1 to 2 feet high, with one or two leaves rising out of a spreading tuft of radical leaves, rather large and ovate, coarsely toothed; flower-heads large and yellow, generally 3 or 4, but sometimes many more, in a loose terminal corymb. Banks, old walls, and rocks about Torquay. High Tor rocks. Rocks near Dunsford, *Fl. D.* (E. B. t. 2082.) P. VI. VII.

3. **H. umbellatum** (*narrow-leaved H.*)—In woods, hedges, and stony places, not uncommon. Root forming buds in the autumn, which do not expand into a tuft of spreading leaves, but in the following summer rise up into a leafy, erect stem, from 1 to 4 feet high, spreading out at the top into an umbel-like corymb of 5 or more flower-heads, bearing yellow florets. Leaves of the flower-stem linear or nearly so, coarsely toothed, nearly sessile. Holne Chase. Ivybridge. Near Dartmouth Castle. Gidleigh, near Chagford. Fingle bridge, on the Teign. (E. B. t. 1771.) P. VII.-IX.

4. **H. boreale** (*shrubby broad-leaved H.*)—In woods and shady places, or under hedges. Plants from 2 to 4 feet high. Stem leafy; leaves oval-lance-shaped, upper ones sessile, lower almost narrowed into a stalk. Flower-heads in a leafy corymb; flowers pale-yellow. Ivybridge. N. Bovey, Hennock, *Fl. D.* (E. B., *H. Subaudum*, t. 349.) P. VIII. IX.

LAPSANA. NIPPLE-WORT.

L. communis (*common N.*)—Common on hedge-banks, waste

CALYCIFLORÆ. 67

places, and roadsides. Plant from 2 to 4 feet high, hairy at the base. Leaves thin and hairy; lower ones ovate, distantly toothed; upper small, narrow and entire. Flower-heads on slender stalks, in a loose corymb; flowers small and yellow. Seed not feathered. Common everywhere. (E. B. t. 844.) A. VII. VIII.

CICHORIUM. CHICORY, SUCCORY.

C. Intybus (*wild Chicory, or S.*)—In dry waste places, roadsides, and borders of fields. Stem from 2 to 3 feet high, covered with bristly hairs, rising from a long tapering root. Lower leaves long and jagged, spreading on the ground; upper leaves oblong, clasping the stem, and with smooth edges. Flower-heads in pairs, sessile between the leaves and the stem; flowers of a pale but bright blue. Seeds smooth and closely packed in the dry involucre. Side of the lane leading to Forde bog, near Newton. Teignmouth. Kingskerswell. Lanes about Paignton. (E. B. t. 539.) P. VII. VIII.

TRIBE II. CYNAROCEPHALÆ. ARTICHOKE OR THISTLE TRIBE.

ARCTIUM. BURDOCK.

A. Lappa (*common B.*)—By roadsides and in waste places, very frequent. A stout, strong, branching plant, from 3 to 5 feet high; the lower leaves large and heart-shaped, stalked; upper leaves much smaller and broadly oval, green and smooth above, but covered with a white cottony down beneath. Flower-heads panicled, the small reddish-purple florets crowning the hairy globular involucres: these last downy and covered with hooked scales which cause them to fasten themselves to one's clothes or the coats of animals. Common everywhere by waysides. (E. B. t. 1228.) B. VIII.

SERRATULA. SAWWORT.

S. tinctoria (*common S.*)—In thickets, open woods, and moist pastures. Plant from 1 to 3 feet high, not much branched; the lower leaves more or less pinnate, with long lance-shaped, pointed, and finely-toothed segments. Flower-heads in a corymb, crowning the stem; flowers purple, the male flowers longer than the female. Involucres small. Ilsham, near Torquay. Marychurch. Bovey Heath. (E. B. t. 38.) P. VIII.

CARDUUS. THISTLE.

1. **C. nutans** (*musk T.*)—In waste places, frequent in the south of England. A stout strong plant, from 2 to 3 feet high; leaves sinuate, deeply cut and spinous. Flower-heads large, and nodding over; florets reddish-purple. Torquay. Marychurch, etc. (E. B. t. 1112.) B. v.-viii.

2. **C. acanthoides** (*welted T.*)—Roadsides and waste places. Resembling the last, but taller and more branched, leaves narrower, and both they and the stem more thickly covered with prickles. Flower-heads smaller and not drooping so much. Flowers deep purple. Marychurch. Watcombe. Chudleigh. Paignton. (E. B. t. 973.) *Carduus crispus*, Bab. B. viii.

3. **C. tenuiflorus** (*slender-flowered T.*)—Sandy places and waste ground, near the sea. Stem from 2 to 4 feet high, having broad, deeply cut, spinous wings throughout its entire length, formed by the decurrent bases of the leaves, which are broadly lanceolate, sinuate, and spinous, their under sides cottony. Flower-heads numerous, crowded together, narrow and sessile; florets pink. Babbicombe Down. Berry Pomeroy Castle. Watcombe. (E. B. t. 412.) A. or B. vi.-viii.

4. **C. Marianus** (*milk T.*)—Sides of cliffs and waste places. Stem from 3 to 5 feet high, ribbed and furrowed, but not winged. Leaves large and broad, oblong-lanceolate, clasping the stem, beautifully variegated with green and milk-white veins. Flower-heads large and globular, with long recurved spines from the scales of the involucre. Florets purple, with long tubes. Rocky Valley. Babbicombe. Watcombe. Chudleigh Rocks. (E. B. t. 976.) *Silybum*, Bab. B. vi. vii.

CNICUS. PLUME-THISTLE.

1. **C. lanceolatus** (*spear Plume-Thistle.*)—Pastures and roadsides. Plant from 3 to 4 feet high; stem upright, and armed with long, pinnatifid, thorny leaves, which are white and downy on their under surface. Heads of flowers single and large; scales of the involucres spear-shaped and spreading; florets purple. Common by waysides, etc. *Carduus*, Linn. (E. B. t. 107.) B. vii. viii.

2. **C. palustris** (*marsh P.*)—In wet meadows, and by moist shady banks. Plant from 4 to 6 feet high; stem clothed with numerous short and sharp spines; leaves pinnatifid, lanceolate, spinous, spreading back on to the stem at their base. Flower-

heads in a thickly aggregated cluster, with ovate involucres; florets purple, sometimes white. Very common in moist situations. *Carduus*, Linn. (E. B. t. 974.) A. VII. VIII.

3. **C. arvensis** (*creeping P.*)—In fields and by roadsides. Plant from 3 to 4 feet high; stem angular, leafy; leaves oblong-lanceolate, spinous, sessile, pinnatifid; flower-heads in an imperfect corymb; florets purple; involucres ovate. Very common everywhere. (*Carduus*, E. B. t. 975.) P. VII.

4. **C. eriophorus** (*woolly-headed P.*)—In waste grounds and by roadsides. Plant from 3 to 4 feet high; stem stout, much branched and furrowed; root-leaves very long, pinnatifid, with lobes pointing alternately up and down, spinous; stem-leaves smaller, but having the same characters, cottony beneath, and half clasping the stem. Flower-heads large and globose; florets purple. Frequent. Watcombe. (*Carduus*, E. B. t. 386.) B. VIII.

5. **C. pratensis** (*meadow P.*)—In wet meadows. Stem from 1 to 2 feet high, cottony, bearing a single head of florets. Leaves nearly all radical, lance-shaped, toothed, and bordered with small, sharp prickles, sessile and cottony beneath. Flower-heads solitary, roundish, slightly cottony; florets purple. Marychurch. Moist meadows, near Torre Abbey. (*Carduus*, E. B. t. 177.) P. VI.–VIII.

6. **C. acaulis** (*dwarf P.*)—In dry limestone pastures. This plant may be said to have no stem, the flower-head arising from the midst of the spreading root-leaves, which are oblong and pinnately divided; the lobes three times cut, and spinous. Flower-head nearly sessile; involucre ovate; florets spreading, purple. Babbicombe and Ilsham Downs, formerly, but not met with for the last two or three years. *Carduus*, Linn. (E. B. t. 161.) P. VII.–IX.

ONOPORDUM. COTTON-THISTLE.

O. Acanthium (*common Cotton-Thistle.*)—In waste grounds and by roadsides. Plant from 4 to 5 feet high, with a branched, woolly and winged stem, bearing sharp spines; leaves ovate-oblong, prickly and woolly on both sides. Involucre large and round; scales armed with sharp teeth. Florets purple. Meadfoot. (E. B. t. 977.) B. VIII.

CARLINA. CARLINE-THISTLE.

C. vulgaris (*common C.*)—In dry fields and hilly pastures. About a foot high, stem woolly with short spines, branching at the top. Radical leaves spear-shaped, spinous; stem-leaves clasping, cottony on their under sides. Flower-heads large; inner scales of the involucre cream-coloured; florets red, with yellow anthers. Fields about Torquay. Stentaway Hill. Babbicombe Down. (E. B. t. 1144.) B. VII.-X.

CENTAUREA. KNAPWEED, BLUE-BOTTLE, AND STAR-THISTLE.

1. **C. nigra** (*black discoid Knapweed.*)—In meadows and by roadsides, frequent. Stem from 1 to 2 feet high, branched; branches bearing one flower-head. Lower leaves stalked, lanceolate, sometimes entire and sometimes deeply notched; upper leaves sessile. Flower-heads large and round; florets purple, the outer row spreading. Hedges and fields, very common. (E. B. t. 278.) P. VIII. IX.

2. **C. scabiosa** (*greater K.*)—Fields, hedges, and banks. From 2 to 3 feet high; stem erect, much-branched, rough and furrowed. Leaves pinnatifid, rather rough; segments lanceolate, pointed. Heads on long, leafless stalks, solitary; involucre large, nearly round, and generally rather woolly. Florets purple, the outer row spreading out in rays around the top of the involucre. Fields about Torquay and Marychurch. Warberry Hill. Meadfoot Cliffs. (E. B. t. 56.) P. VII.-IX.

3. **C. Cyanus** (*corn Bluebottle.*)—In cornfields, common. Stem from 1 to 3 feet high, branched, cottony. Leaves numerous, linear-lanceolate; root-leaves slightly toothed; both stem and leaves downy. Flower-heads solitary; involucre oblong; inner florets purple, outer row large and spreading, bright-blue. Cornfields, common. (E. B. t. 277.) A. VI.-VIII.

4. **C. Calcitrapa** (*common Star-Thistle.*)—Gravelly and sandy places. About a foot high; stem branched, furrowed and slightly hairy. Root-leaves deeply pinnatifid, with lanceolate segments; stem-leaves not so deeply cut, with linear lobes. Flower-heads sessile, the scales of the involucre ending in long sharp points; florets purple. Exmouth Sands, *Fl. D.* (E. B. t. 125.) A. VII.-IX.

TRIBE III. CORYMBIFERÆ.

SUBTRIBE 1. TUBIFLORÆ.

BIDENS. BUR-MARIGOLD.

1. **B. cernua** (*nodding B.*)—Sides of ditches and in marshy

CALYCIFLORÆ. 71

places. Plant from 1 to 3 feet high; stem branching. Leaves clasping the stem, lanceolate, coarsely serrated; flower-heads terminal, drooping; involucre surrounded by long lanceolate bracteas. Florets of a dull greenish-yellow. Goodrington Marsh. (E. B. t. 1114.) A. VII.–IX.

2. **B. tripartita** (*trifid B.*)—In marshy places, by sides of ponds and lakes. From 1 to 3 feet high. Leaves on short stalks, and divided into three lanceolate and serrated segments. Flower-heads terminal, smaller than in the last, slightly drooping; florets brownish-yellow. Ditches about Chudleigh, *Fl. D.* (E. B. t. 1113.) A. VII.–IX.

TANACETUM. TANSY.

T. vulgare (*common Tansy.*)—Under hedges, by roadsides, and in waste places. Plant from 1 to 3 feet high; stem erect, not much branched, very leafy. Leaves twice pinnatifid, with serrated segments. Flower-heads in a terminal corymb; flowers bright golden-yellow. Involucre hemispherical. Babbicombe. Newton road, between Kingskerswell and Newton. Lane by the side of the cricket-ground at Highweek. (E. B. t. 1229.) P. VII. VIII.

ARTEMISIA. WORMWOOD, SOUTHERNWOOD, MUGWORT.

1. **A. vulgaris** (*common Mugwort.*)—By sides of hedges and in waste places, common. Plant from 2 to 3 feet high, with a short, woody, and thick stock, throwing off erect flowering stems; leaves deeply pinnatifid, with lanceolated, coarsely-toothed segments, green on their upper, but very white on their under surfaces. Flower-heads in a long terminal cluster; involucres ovoid and cottony; florets reddish or brownish-yellow, from 12 to 20 in each head. Torre Abbey, and waste places near Torquay. Marychurch. Teignmouth, etc. (E. B. t. 978.) P. VII.–IX.

2. **A. Absinthium** (*common Wormwood.*)—By roadsides and in waste places, frequent. Growing from 1 to 3 feet in height; stock short, branched, and leafy, with erect and hard flowering stems; whole plant covered with a soft silky down, which gives it a greyish-white appearance. Leaves, in their general contour, almost round, but very much divided into linear-oblong segments; uppermost leaves oblong, nearly entire. Flower-heads numerous, in terminal, leafy clusters, drooping, nearly round;

florets of a dingy-yellow colour. Near Hope's Nose. (E. B. t. 1230.) P. VIII. IX.

3. **A. maritima** (*sea Wormwood.*)—On the seashore and in salt-marshes. A very much branched, decumbent plant, more or less covered with a thick, white, cottony down. Leaves twice pinnate; segments nearly linear. Flower-heads in drooping clusters, oblong; florets from 3 to 6, pale dingy-yellow. The variety *A. Gallica* (E. B. t. 1001) is frequently found growing side by side with this. Paignton. Goodrington Marsh. Teignmouth, on the waste ground under the sea-wall. (E. B. t. 1706.) P. VIII. IX.

EUPATORIUM. HEMP-AGRIMONY.

E. cannabinum (*common H.*)—Sides of rivers and ditches. Stems numerous, from 2 to 4 feet high, branched, downy. Leaves on very short stalks, opposite, downy, 3 or 5 times divided; leaf-segments lanceolate and deeply notched. Flowers very numerous, in a densely crowded corymb, pale reddish-purple. Ansti's Cove lane. Side of the Paignton road. (E. B. t. 428.) P. VII.-IX.

LINOSYRIS. GOLDILOCKS.

L. vulgaris (*flax-leaved G.*)—On limestone cliffs, very local. Plant from 6 inches to more than a foot high; stem erect, crowded with numerous narrow-linear leaves. Flower-heads in a thickly aggregated terminal corymb; florets of a bright golden yellow. Berry Head, plentiful. *Chrysocoma Linosyris*, Linn. (E. B. t. 2505.) A. VIII. IX.

GNAPHALIUM. CUDWEED.

G. uliginosum (*marsh C.*)—In wet, sandy fields. Plant about 6 or 7 inches high, much branched, cottony. Leaves narrow-oblong or linear, few. Flower-heads in small clusters within the tufts of leaves at the extremities of the branches; florets of a shining yellowish-brown. Paignton. Ilsham. Goodrington. (E. B. t. 1194.) A. VII.-IX.

FILAGO. FILAGO.

1. **F. minima** (*least Filago.*)—In dry and gravelly places.

CALYCIFLORÆ. 73

A very slender plant, with an erect, cottony stem, irregularly branched at the top. Leaves also cottony, linear-lanceolate, and pressed up toward the stem. Flower-heads small and numerous, in little axillary bunches; florets of a pale-yellow. Exmouth Warren. Middledon Down, near Chagford. (*Gnaphalium*, E. B. t. 1157.) A. VI.-IX.

2. **F. Germanica** (*common F.*)—In dry fields and stony or sandy waste places. Stem from 6 to 8 inches high, with numerous upright, cottony, lanceolate or linear leaves; a roundish cluster of flower-heads crowns the main stems, from beneath which 2 or 3 branches radiate, each bearing a similar cluster of flowers; florets pale-yellow. Very common. Park Hill. Warberry Hill, etc. (*Gnaphalium*, E. B. t. 946.) A. VII.-IX.

PETASITES. BUTTER-BUR.

P. vulgaris (*common B.*)—In wet meadows and by roadsides. This plant has an extensively creeping root, sometimes extending for many yards. Leaves very large, spreading on the ground, roundish heart-shaped, unequally toothed, and downy beneath. Flowers, which appear before the leaves, on stems from 4 to 8 inches high, in numerous flower-heads of a dull flesh-colour. The sweet Coltsfoot (*Tussilago fragrans*) grows also very plentifully about Torquay. *Tussilago Petasites*, Linn. (E. B. t. 431.) P. III.-V.

SUBTRIBE 2. RADIATÆ.

TUSSILAGO. COLTSFOOT.

T. Farfara (*Coltsfoot.*)—In moist clayey waste and cultivated ground. Roots creeping. Leaves roundish heart-shaped, angular, toothed, downy on the under surface and slightly so on the upper; flowers appear before the leaves in solitary heads with florets of a bright yellow colour. Fields and roadsides, too common. (E. B. t. 429.) P. III. IV.

ASTER. STARWORT, MICHAELMAS-DAISY.

A. Tripolium (*sea S., or M.*)—In salt-marshes and muddy banks of tidal rivers. Stem about 1 foot high, slightly branched. Leaves succulent, sessile, linear-lanceolate, not hairy. Flower-heads in a close corymb; outer florets bluish-purple, inner florets yellow. Banks of the Dart, near Sharpham. (E. B. t. 87.) P. VIII. IX.

SOLIDAGO. GOLDEN-ROD.

S. Virgaurea (*common G.*)—In woods and thickets. Plant from 6 inches to 2 feet or more high; stem upright, rigid, generally slightly downy. Lower leaves stalked, ovate, ending in a point, slightly serrate: stem-leaves lanceolate, tapering towards the stem, but not stalked. Flower-heads in a crowded panicle; florets bright yellow, outer ones radiating. Park Hill, near the Quarry. Warberry Hill. Petit Tor. (E. B. t. 301.) P. VII.-IX.

SENECIO. GROUNDSEL, RAGWORT, FLEAWORT.

1. **S. vulgaris** (*common Groundsel.*)—By sides of roads, in fields and waste grounds. From 3 or 4 inches to more than a foot high. Leaves half clasping the stem, pinnatifid, and unevenly toothed. Flower-heads in a loose corymb, small, with minute yellow florets. Abundant. (E. B. t. 747.) A. I.-XII.

2. **S. sylvaticus** (*mountain G.*)—On dry banks, in waste places and outskirts of woods. A taller and weaker plant than the last, slightly downy; leaves sometimes clasping and auricled at the base. Flower-heads corymbose; florets small and yellow. Meadfoot. Hope's Nose, near the raised beach. (E. B. t. 748.) A. VII.-IX.

3. **S. tenuiflorus** (*hoary Ragwort.*)—Under hedges and by roadsides. Stem erect, about 2 feet high, slightly cottony; leaves pinnatifid; segments narrow, clasping, downy and white on their under surface; lower leaves stalked. Flower-heads numerous, in a loose cluster; florets yellow, outer ray broad and spreading. Upton. Marychurch. *Senecio erucæfolius*, Bab. (E. B. t. 574.) P. VII. VIII.

4. **S. Jacobæa** (*common R.*)—Waste grounds, pastures, and roadsides. Stem erect, branched, from 2 to 3 feet high, leafy. Lower leaves stalked, of a long-oval shape, lyrate; upper leaves twice pinnatifid, sessile, with oblong deeply-toothed segments. Heads of flowers large, in upright corymbs, golden-yellow, with an outer ray of spreading florets. Very plentiful everywhere. (E. B. t. 1130.) P. VII.-IX.

5. **S. aquaticus** (*marsh R.*)—In marshy places and by rivers and ditches. Stem from 1 to 4 feet high, branched in its upper

half. Lower leaves obovate, entire, and stalked; upper leaves lyrate and cut into oblong or linear segments. Heads of flowers in corymbs, similar to but larger than those of *S. Jacobæa.* Forde bog, near Newton. (E. B. t. 1131.) P. VII. VIII.

INULA. INULA.

1. **I. Helenium** (*Elecampane.*)—In moist pastures, rare. Stem from 3 to 4 feet high, strong, round, and furrowed, branching above. Leaves large, oval heart-shaped, clasping, slightly serrate, white and downy beneath; root-leaves stalked. Flower-heads large, with ovate, leafy, and spreading involucre-scales; florets bright golden-yellow. Orchards at Rora, near Ilsington, *Fl. D.* (E. B. t. 1546.) P. VII. VIII.

2. **I. Conyza** (*Ploughman's Spikenard.*)—On dry, chalky or limestone soils. Stem erect, 2 or 3 feet high, angular. Leaves hairy, broadly lanceolate, serrate; lower leaves stalked. Flower-heads in a leafy cluster at the top of the stem; florets yellow, those of the circumference very small. Park Hill. Meadfoot Cliffs. Warberry Hill. (*Conyza squarrosa,* E. B. t. 1195.) P. VII.-IX.

PULICARIA. FLEA-BANE.

P. dysenterica (*common F.*)—In damp situations. Stem from 1 foot to 18 inches high, branched and woolly. Leaves downy, alternate, clasping, oblong and pointed at the ends, their margins waved and slightly indented. Flower-heads in a loose terminal panicle, some axillary; flowers numerous, of a rich bright yellow, rays much longer than the disc. Goodrington. Teignmouth. Exmouth. *Inula,* Linn. (E. B. t. 1115.) P. VII.-IX.

BELLIS. DAISY.

B. perennis (*common D.*)—Pastures, very abundant. Flower-stalks arising from the midst of a cluster of spreading, oblong, crenated leaves. Flower-heads solitary; florets of the ray white, tinged with pink at the extremities, those of the disk yellow. The "wee, modest, crimson-tipped flower" of [Burns, which glistens, star-like, over all our meadows and pastures. (E. B. t. 424.) P. II.-X.

CHRYSANTHEMUM. OX-EYE.

1. **C. Leucanthemum** (*great white Ox-eye.*)—In fields and dry pastures. Stem erect, branched, from 1 to 2 feet high. Lower leaves stalked, obovate, coarsely toothed; stem-leaves narrow, oblong, pinnatifid at their junction with the stem, sessile. Flowers in solitary heads, large; florets of the ray white, long, and spreading, those of the centre numerous and yellow. Very common. (E. B. t. 601.) A. VI. VII.

2. **C. segetum** (*corn Marigold, yellow Ox-eye.*)—In cultivated lands and cornfields. Stem from 6 inches to a foot high, erect, branching at the top. Lower leaves stalked; upper ones narrow and clasping, irregularly serrated, much more succulent in appearance than the last. Flower-heads large, solitary, bright yellow; florets of the ray broad and flat. Warberry Hill. Chudleigh. Exmouth. (E. B. t. 540.) A. VI.-VIII.

MATRICARIA. WILD-CHAMOMILE, FEVERFEW.

1. **M. inodora** (*corn W., or scentless Mayweed.*)—In fields and by waysides. Stem from 12 to 18 inches high, erect, branched. Leaves sessile, twice pinnate, segments very narrow, pointed. Flowers on long stalks, terminal, solitary; ray of long white florets; disk yellow, convex. Fields and pastures about Torquay. (*Pyrethrum*, E. B. t. 676.) The variety β of Hooker and Arnott, *M. maritima*, with fleshy leaves and hemispherical receptacle, grows on Meadfoot shore and Paignton sands. (*Pyrethrum*, E. B. t. 979.) A. VI.-XI.

2. **M. Chamomilla** (*wild Chamomile.*)—In fields and waste places. Stem erect, about a foot high. Leaves twice or three times pinnate, with short but narrow-linear segments. Flower-heads large, on terminal flower-stalks; ray composed of white oblong florets; disk prominently conical. Very common in the vicinity of Torquay, etc. (E. B. t. 1232.) A. VI.-VIII.

3. **M. Parthenium** (*common Feverfew.*)—By roadsides and in waste places. Stems erect, branching, a foot or more high. Leaves stalked and pinnate; segments oblong, lobed and cut, of a dull green colour. Flower-heads corymbose; florets of the ray short and white, those of the disk yellow and numerous. Kingskerswell. (E. B. t. 1231.) P. VII.-IX.

ANTHEMIS. CHAMOMILE.

1. **A. nobilis** (*common C.*)—In gravelly pastures and sandy waste places. Stem about a foot high, drooping, and much branched. Leaves twice pinnate, segments linear and pointed. Flower-heads at the end of every branch, solitary, with white rays and yellow disks, which latter become conical as the flowering advances. Marychurch. Milber Down. Bovey Heathfield. (E. B. t. 980.) P. VII.–IX.

2. **A. arvensis** (*corn C.*)—Cornfields, and by hedgerows in cultivated fields. Stem from 1 to 2 feet high, erect, much branched, hoary. Leaves twice pinnate, segments linear-lanceolate, hairy. Flower-heads on long stalks, terminal and solitary. Florets of ray large and white, those of the disk small and yellow. Moreton, Ilsington, *Fl. D.* (E. B. t. 602.) A. VI.–VIII.

3. **A. Cotula** (*stinking C.*)—In fields and waste places. Stem erect, branching, from 10 to 18 inches high. Lower leaves 2 or 3 times, upper ones once pinnate; segments short, but narrow and pointed. Flowers in solitary terminal heads, disk convex and yellow, ray large and white. Stem sprinkled with glandular dots. (E. B. t. 1772.) A. VI.–IX.

ACHILLEA. YARROW, MILFOIL.

1. **A. Ptarmica** (*Sneezewort Yarrow.*)—In meadows and pastures. Stem 18 inches or 2 feet high, upright, leafy, slightly branched in its upper part. Leaves linear-lanceolate, pointed, minutely serrated, shining. Flower-heads in a large and handsome corymb; florets of both disk and ray white. Road to Barton Ridge. Banks of the Teign. Chudleigh. (E. B. t. 757.) P. VI.–VIII.

2. **A. Millefolium** (*common Y., or Milfoil.*)—In pastures, waste places, and by roadsides, common. Plant from 6 inches to nearly 2 feet high; stem erect, branching at the top. Leaves finely cut into a multitude of short but extremely fine and deeply pinnated segments. Flower-heads numerous, small, crowded together into a thick terminal corymb; flowers white or pinkish. Both leaves and stem sometimes rather woolly. (E. B. t. 758.) P. VI.–IX.

ORD. XLVII. CAMPANULACEÆ.

CAMPANULA. BELL-FLOWER.

1. **C. rotundifolia** (*round-leaved B., or Harebell.*)—On hilly

and dry pastures. Stem from 6 to 12 inches high, very slender. Radical leaves roundish heart-shaped, on long stalks; lower leaves of the stem lanceolate, stalked; upper ones linear and sessile. Flowers one or more on a stalk, bell-shaped, 5-cleft at the mouth, of a beautiful blue colour, and gracefully bending over. Berry Head. Woodhouse lane, near Ilsington, *Fl. D.* (E. B. t. 866.) P. VII. VIII.

2. **C. hederacea** (*ivy-leaved B.*)—In marshy places and peaty bogs. Stem prostrate and thread-like; leaves roundish, cut into 5 angular lobes, of a delicate transparent green; flowers on long stalks, pale reddish-blue, at first nodding over, but when fully open erect. Holne Chase. Foot of Middledon Down, near Chagford. Ivybridge. Bovey Heath. Forde bog, near Newton. Haldon. *Wahlenbergia*, Bab. (E. B. t. 73.) P. VII. VIII.

3. **C. hybrida** (*corn B.*)—In cornfields. Stem from 6 to 12 inches high, covered with small stiff hairs, sparingly branched. Leaves oblong, wavy at the edges, sessile; lower leaves tapering towards the stem. Flowers few, solitary, reddish-purple, small, surrounded by the spreading divisions of the calyx, and surmounting the long, triangular seed-vessel. Fields above Meadfoot. Cliff-walks between Meadfoot and Hope's Nose. Cornfields at Marychurch. *Specularia*, Bab. (E. B. t. 375.) A. VI.-IX.

JASIONE. SHEEP'S-BIT.

J. montana (*annual S.*, or *Scabious.*)—On dry, heathy pastures. Stems from 6 to 10 inches high, branched, arising from the crown of the root, surrounded by numerous, spreading, bluntish-oblong leaves, which are rough and wavy. Flowers light but bright blue, in closely crowded terminal heads, surrounded by a leafy involucre. Meadfoot Cliffs. Warberry Hill, etc. (E. B. t. 882.) B. VI.-IX.

ORD. XLVIII. **LOBELIACEÆ.**

ORD. XLIX. **VACCINIACEÆ.**

VACCINIUM. WHORTLEBERRY.

V. Myrtillus (*Bilberry*, or *W.*)—In woods and heaths in rocky districts. A stiff, woody plant, from 1 to 2 feet high.

Stem branched, sharply angular. Leaves ovate, on short stalks, smooth, with serrated margins. Flowers solitary, drooping, light red, with a greenish tinge. Berries bluish-black. The berries are very pleasant to the taste, and made into tarts and served up with clotted cream would be pronounced delicious by the most fastidious epicure. Plantations at Lindridge, near Bishop's Teignton. Holne Chase. Ivybridge. Moor near Chagford. (E. B. t. 456.) Sh. v.

SUB-CLASS III. COROLLIFLORÆ. (ORD. L.-LXVIII.)

A. Stamens free from the Corolla. (Ord. L.-LII.)

ORD. L. ERICACEÆ.

ERICA. HEATH.

1. **E. Tetralix** (*cross-leaved H.*)—On heaths, downs, and moors, common. Stem branched in its lower part and very leafy. Leaves 4 in a whorl, linear or lanceolate, downy above, becoming more distant towards the extremities of the twigs, and leaving the space beneath the flowers bare. Flowers rose-coloured, in a terminal drooping cluster. Milber Down. Forde bog, near Newton. Bovey Heath, etc. (E. B. t. 1014.) Sh. VII. VIII.

2. **E. cinerea** (*fine-leaved H.*)—On dry heaths and downs, abundant. Plant with many upright stems. Leaves 3 in a whorl, linear-lanceolate, flat above, with generally little bundles of small leaves in their axils. Flowers in long whorled clusters, reddish-purple, drooping. Warberry Hill. Babbicombe. Marychurch. Milber Down (with the white variety). (E. B. t. 1015.) Sh. VII. VIII.

CALLUNA. LING.

C. vulgaris (*common L.*)—On dry heaths and moors, very common. Low and straggling, seldom more than 1 foot high. Leaves small and opposite, prolonged slightly at the base; flowers small, of a bluish-pink, drooping, sometimes approaching to or quite white. Babbicombe Down. Milber Down. Bovey Heath, etc. (E. B. t. 1013.) Sh. VI.-VIII.

Ord. LI. PYROLACEÆ.

Ord. LII. MONOTROPACEÆ.

B. Stamens inserted upon the Corolla. (Ord. LIII.–LXVIII.)

Ord. LIII. AQUIFOLIACEÆ.

ILEX. HOLLY.

I. Aquifolium (*common H.*)—In hedges and woods, frequent. A small tree, sometimes not above a shrub. Leaves evergreen, ovate and pointed, wavy, with a shining surface, and strong spinous teeth at their edges; those of the upper branches often entire and without spines. Flowers white, on short axillary stalks, in thick clusters. Berries bright red. Common in hedges and woods around Torquay, etc. (E. B. t. 496.) T. vi.–viii.

Ord. LIV. OLEACEÆ.

LIGUSTRUM. PRIVET.

L. vulgare (*Privet.*)—In hedges and thickets, very frequent. A shrub growing from 6 to 8 or 10 feet high. Leaves small and numerous, on short stalks, oblong or lanceolate. Flowers small and white, in short thick clusters at the ends of the branches. Berries round and black, sometimes, though rarely, yellow. Abundant. Park Hill. Ansti's Cove, etc. (E. B. t. 764.) Sh. vi. vii.

FRAXINUS. ASH.

F. excelsior (*common A.*)—Woods and hedges. One of our most handsome trees. Leaves pinnate, with from 4 to 8 pairs of broadly lanceolate leaflets. Flowers, which are without calyx or corolla, appear before the leaves. Common. (E. B. t. 1692.) T. iv. v.

Ord. LV. APOCYNACEÆ.

VINCA. PERIWINKLE.

1. **V. minor** (*lesser Periwinkle.*)—On shady banks and in woods. Stem procumbent, leaves narrow ovate or lanceolate, flowering branches erect. Flowers blue, salver-shaped, solitary, on axillary flower-stalks. Cockington. Coffin's Well. (Blagdon, *Mr. Earle.*) Chudleigh. White variety, bottom of Bunker's Hill, near Totness, *Fl. D.* (E. B. t. 917.) P. IV.–VI.

2. **V. major** (*greater P.*)—In woods and thickets. Much larger than the former; leaves oval heart-shaped. Flowers large, purplish-blue. Met with about Torquay occasionally, but probably a truant from shrubberies. Near Ide, *Fl. D.* (E. B. t. 514.) P. IV.–V.

Ord. LVI. GENTIANACEÆ.

ERYTHRÆA. CENTAURY.

1. **E. Centaurium** (*common Centaury.*)—In dry pastures. Stem from 3 inches to a foot high, square, branched above. Leaves of an oval-oblong, upper ones pointed; flowers nearly sessile, in a corymbose cluster, rose-coloured. Daddy-hole Plain. Babbicombe, etc. (E. B. t. 417.) A. VI.–IX.

2. **E. pulchella** (*dwarf branched C.*)—Sandy ground, generally near the sea. Stem from 2 to 6 inches high, much branched; leaves ovate-oblong. Flowers in a loose leafy cluster, pink. Chapel Hill. Babbicombe. Petit Tor. (E. B. t. 458.) A. VII.–IX.

GENTIANA. GENTIAN.

1. **G. Amarella** (*small-flowered Gentian.*)—In dry limestone pastures. Stem erect, square, from 3 to 12 inches high, but extremely variable in size. Leaves sessile, lower ones oval, upper oblong-lanceolate. Flowers pale dull-purple, corolla 5-cleft, and calyx with five lobes. Babbicombe Down. (E. B. t. 236.) A. VIII.–IX.

2. **G. campestris** (*field G.*)—Similar situations to the last. Stem from 3 to 10 inches long, very much branched; leaves elliptical. Flowers pale sickly-blue, corolla 4-cleft, calyx with four lobes. Babbicombe Down. Watcombe. (E. B. t. 237.) A. VIII.–IX.

CHLORA. YELLOW-WORT.

C. perfoliata (*perfoliate Yellow-wort.*)—In chalky and limestone pastures. Stem from 1 foot to 18 inches high, simple; leaves distant, in pairs, connected at the base by their whole breadth, and pierced by the stem, triangularly ovate; stem branching at the top into a flowering panicle, bearing bright yellow flowers; corolla and calyx both consisting of 8 divisions. Near Starcross, *Fl. D.* (E. B. t. 60.) A. VII.–IX.

MENYANTHES. BUCKBEAN.

M. trifoliata (*Buckbean, or marsh Trefoil.*)—In marshy and boggy situations. Root creeping; stem thick, round, and leafy; leaves stalked, divided into three equal, obovate, wavy leaflets; flower-stalk erect, terminating in a dense cluster of flesh-coloured flowers, thickly fringed with beautiful white filaments. Marshy meadow at Edginswell. Forde bog, near Newton. Bovey Heath. (E. B. t. 495.) P. V.–VII.

Ord. LVII. **POLEMONIACEÆ.**

Ord. LVIII. **CONVOLVULACEÆ.**

CONVOLVULUS. BINDWEED.

C. arvensis (*small Bindweed.*)—In fields, hedges, and waste places. Stems twining or decumbent, leafy, numerous; leaves alternate, arrow-shaped, with acute lobes, stalked; flower-stalks generally 1-flowered; flowers bell-shaped, variegated with rose-colour and white. Very common in all our fields and hedges. (E. B. t. 312.) P. VI.–VIII.

CALYSTEGIA. HOODED BINDWEED.

1. **C. sepium** (*great Convolvulus, or hooded Bindweed.*)—In shady woods and hedges. Stem climbing, with large but distant leaves, alternate, arrow-shaped, their lobes truncated. Flowers large and handsome, pure white, sometimes streaked with pink.

Ansti's Cove. Hedges by side of the Paignton road. *Convolvulus,* Linn. (E. B. t. 313.) P. VI.-VIII.

2. **C. Soldanella** (*seaside C.*)—On sandy seashores. Stem short and prostrate; leaves kidney-shaped, fleshy, stalked. Flowers few, pink, bell-shaped, on thick angular flower-stalks. Paignton sands, near the garden wall of Torbay House. Goodrington sands. Teignmouth. Exmouth Warren. *Convolvulus*, Linn. (E. B. t. 314.) P. VI.-VIII.

CUSCUTA. DODDER.

C. Epithymum (*lesser D.*)—On furze, heath, and thyme. Plant consisting of small, red, thread-like stems, twisting and tangling about the branches of the plant upon which it is parasitic, destitute of leaves. Flowers sessile, crowded, reddish-white; corolla tubular, divided into four ovate pointed segments. Warberry hill, in the furze-brake at the summit. Near Teignmouth. Bovey Heath. Dawlish. (E. B. t. 55.) A. VII.-X.

ORD. LIX. BORAGINACEÆ.

ECHIUM. VIPER'S BUGLOSS.

E. vulgare (*common Viper's Bugloss.*)—On banks and old walls, in fields and waste ground, very common. A very handsome, showy plant. Stem from 1 to 3 feet high, strong, round, sprinkled with red hairy tubercles. Root-leaves long and lanceolate, hairy, stalked; upper leaves nearly sessile. Flowers in short spikes from the sides of the stem, closely crowded in the upper part, at first a rich red-purple, afterwards bright blue. Warberry Hill. Down between Meadfoot and Hope's Nose. Aller. (E. B. t. 181.) B. VI. VII.

LITHOSPERMUM. GROMWELL.

1. **L. officinale** (*common Gromwell.*)—In dry waste places, and uncultivated, rubbishy spots. Plant from 1 to 2 feet high, much branched; stem covered with stiff hairs; leaves alternate, lanceolate, hairy beneath; flowers small, pale yellow; seeds polished, whitish-brown, very hard. Very frequent about Torquay. Marychurch, etc. (E. B. t. 134.) P. V.-VII.

2. **L. arvense** (*corn G., or bastard Alkanet.*)—Occasionally

found in cornfields and waste grounds. Not so large and bold as the last; stem upright and branched; leaves lanceolate, hairy; flowers white. Calyx-segments, long and open when the plant is in fruit, surrounding the brown, wrinkled nuts. Barton. Road between Teignmouth and Dawlish. (E. B. t. 123.) A. V. VI.

3. **L. purpureo-cæruleum** (*creeping or purple G.*)—Bushy places on a limestone soil. Rare. Barren stems prostrate, flowering stems erect, from 12 to 18 inches high; leaves alternate, hairy, lanceolate. Flower-heads in a leafy spike; flowers large, of a deep rich blue, with the tube of the corolla reddish, longer than the calyx. Petit Tor, near Marychurch, in the rough, bushy places under Dungeon Cliff. (E. B. t. 117.) P. V.-VII.

MYOSOTIS. SCORPION-GRASS.

1. **M. palustris** (*creeping water S., or Forget-me-not.*)—By sides of streams and in ditches. Root creeping; stem about 12 or 14 inches high. Leaves oblong, bluntly pointed; flower-stems leafless; flowers large and handsome, bright blue with a yellow centre, unopened buds reddish. Banks of streams, etc., common. Forde bog, near Newton, where also *M. repens* is to be found. (E. B. t. 1973, and E. B. S. t. 2703.) P. VI.-VIII.

2. **cæspitosa** (*tufted water S.*)—In watery and boggy places. Root fibrous, not creeping. Stem upright, round; leaves longer and narrower than the last; flower-clusters somewhat leafy below; flowers also smaller, bright blue with a yellow eye. Forde bog, near Newton. (E. B. S. t. 2661.) P. VI.-VIII.

3. **M. sylvatica** (*upright wood S.*)—In shady places and woods. Stems from 12 to 18 inches high, covered with soft spreading hairs; leaves oblong-lanceolate, lower ones on short dilated stalks; flower-stalks diverging, leafless; flowers large, pale blue, very handsome; calyx deeply divided. Woods, frequent, *Fl. D.* (E. B. S. t. 2630.) P. V.-VIII.

4. **M. arvensis** (*field S.*)—In fields and waste places, very frequent. Plant from 6 inches to a foot high; stem branched; leaves oblong, pointed; lower leaves oblong, blunt. Flower-clusters long, bearing numerous small blue flowers. Very common. (E. B. S. t. 2629.) A. VII. VIII.

5. **M. collina** (*early field S.*)—On dry banks and tops of walls. Stem from 4 to 8 inches high, branched, hairy. Leaves oblong, obtuse, lower ones obovate, covered with straight, soft hairs; clusters stalked, with generally one distant flower at the base. Flowers very small, blue. Daddy-hole Plain. Meadfoot

cliffs. Cockington lanes. Upton. Dry hedge-banks on the side of Paignton road, etc. (E. B. t. 2558.) A. III.-V. *M. stricta* of Link, with sessile clusters, Petit Tor, April 1859.

6. **M. versicolor** (*yellow and blue S.*)—In both dry and wet meadows and on banks. Stem from 3 to 12 inches high. Leaves oblong, narrow, pointed, upper ones opposite; clusters stalked; flowers small, yellow before fully expanded, afterwards bright blue. Hope's Nose. Meadfoot cliffs. Forde bog, near Newton. (E. B. t. 480.) A. IV.-VI.

ANCHUSA. ALKANET.

A. sempervirens (*evergreen Alkanet.*)—In waste grounds, stony places, and among ruins. Stem from 1 to 2 feet high, hairy; leaves ovate, pointed, hairy; the lower leaves large and on long stalks; flower-stalks axillary; flowers bright blue. Berry Pomeroy, near the Castle. Stoke Gabriel, in a hedge by the way to the river. Near Ide churchyard. Dartmouth, side of the walk leading to the Castle. (E. B. t. 45.) P. V.-VII.

LYCOPSIS. BUGLOSS.

L. arvensis (*small Bugloss.*)—Cornfields and under hedges. Stem from 12 to 18 inches high, upright, more or less branched towards the top, angular and leafy; leaves lanceolate, with wavy margins, very hairy, sessile; lower leaves tapering into a footstalk. Flowers in a leafy raceme, small, bright blue. Fields near the Parsonage at Moreton. North Bovey, *Fl. D.* (E. B. t. 938.) A. VI. VII.

SYMPHYTUM. COMFREY.

S. officinale (*common Comfrey.*)—In wet places and by the sides of streams. Plant from 18 inches to 4 feet high, rough and hairy, winged in its upper part, branching; root-leaves on long stalks, rough, ovate; upper leaves sessile, lanceolate, wavy. Flower-clusters in pairs, drooping; flowers yellowish-white, sometimes purple. Formerly at Meadfoot, but now extinct. Forde, near Newton. Side of a brook between Kingskerswell and Newton. Exminster. (E. B. t. 817.) P. V. VI

BORAGO. BORAGE.

B. officinalis (*common B.*)—In waste ground and among rubbish. Plant from 12 to 18 inches high; stem erect with spreading branches; lower leaves oblong, pointed, narrowing at the base into long stalks, but clasping the stem by an auricled expansion; uppermost leaves nearly sessile. Whole plant very hairy. Flowers on long stalks, drooping, of a clear brilliant blue, with the dark anthers showing very distinctly in the centre. Paignton. Chudleigh. Bovey. Chagford. (E. B. t. 36.) B. VI. VII.

CYNOGLOSSUM. HOUND'S-TONGUE.

C. officinale (*common H.*)—By roadsides and in waste places. Stem about 2 feet high, stout, erect, and branched, roughly hairy. Radical leaves very long, stalked, ovate-lanceolate, downy; upper leaves lanceolate, half-clasping; flower-clusters numerous, forming a terminal leafy panicle. Flowers small, of a dull purplish-red; seeds covered with short hooked prickles. Paignton Marsh. Near Stoke Gabriel, by the side of the high-road. Kerswell Down, in great abundance. Chudleigh. (E. B. t. 921.) B. VI. VII.

Ord. LX. SOLANACEÆ.

HYOSCYAMUS. HENBANE.

H. niger (*common Henbane.*)—Waste places, in a chalky or limestone soil. A strong, upright, branching plant, from 1 to 2 feet high, hairy and very viscid, with a peculiarly unpleasant odour. Leaves large, sessile, alternate, sharply notched; upper ones clasping, ovoid, and irregularly pinnatifid. Flowers erect, on very short stalks, or sessile; calyx 5-cleft, veined, crowning the fruit after the corolla has fallen; corolla dingy-yellow, most beautifully pencilled with a network of purple streaks; seed-vessel globular, surmounted by the calyx, containing numerous small seeds, and closed by a lid. Whole plant poisonous, with the exception of the seeds. Babbicombe, on the side of the slope at the end of the Down nearest to Petit Tor. Paignton, near the green. Goodrington sands. Chudleigh. (E. B. t. 591.) A. or B. V.–VII.

SOLANUM. NIGHTSHADE.

1. **S. Dulcamara** (*woody N., or Bittersweet.*)—In woods and

hedges. Plant with a woody stem, throwing off climbing and straggling branches; lower leaves heart-shaped, upper ones auricled. Flower-clusters terminal, or opposite to the leaves, drooping; corolla purple, with two green dots at the base of each of the five segments; anthers united in a conical form, large and yellow. Berries oval, scarlet, poisonous. Hedges near Torre Abbey, etc. (E. B. t. 565.) P. VI.-VIII.

2. **S. nigrum** (*common Nightshade.*)—Sides of fields, and waste places. Plant about 1 foot high, with extensively spreading branches. Leaves alternate, ovate, wavy at the edges, on short stalks. Flowers in small lateral clusters, alternate with the leaves, white; berries small, round, and black. Plant poisonous. Near Torre Abbey. Paignton, near the sands. Goodrington sands. (E. B. t. 566.) A. VI.-IX.

Ord. LXI. OROBANCHACEÆ.

OROBANCHE. BROOM-RAPE.

1. **O. major** (*greater Broom-rape.*)—Parasitic on roots of Furze and Broom. Plant from 12 to 18 inches high, leafless, at first of a pale yellow, but afterwards of a lurid purplish-brown; stem thick and fleshy, with small lanceolate scales. Flowers sessile, corolla tubular, upper lip undivided, lower one 3-lobed, the middle one being usually the largest; calyx consisting of from 2 to 4 lanceolate lobes. Daddyhole l'lain, on the descent to the Quarry. Cliffs between Meadfoot and Hope's Nose. Babbicombe. (E. B. t. 421.) *O. Rapum*, Bab. P. V.-VII.

2. **O. minor** (*lesser B.*)—Parasitic on many different kinds of plants, but usually upon Clover. Smaller and more slender in all its parts than the last; stem rather wavy, scales distant. Flower-spike loose, flowers as well as the upper part of the stem light bluish-purple, tube of the corolla contracted in the middle, lobes of the lower lip almost equal, lower part of the stamens hairy. Ide, near Exeter, *Mr. Earle.* (E. B. t. 422.) A. VI.-VIII.

3. **O. Hederæ** (*Ivy B.*)—Parasitic on roots of Ivy. Considered by Bentham to be a variety of the last, which it very much resembles, differing in having the stigma yellow instead of purple. Rock Walk. Daddyhole Plain. (E. B. S. t. 2859.) P. VI.-VIII.

LATHRÆA. TOOTHWORT.

L. Squamaria (*greater Toothwort.*)—In woods and thickets, parasitic on roots of Hazels and other trees. Stem of the plant flesh-coloured; root fleshy and creeping, covered with closely packed, thick scales, resembling somewhat in shape human incisor teeth; flower-stem from 4 inches to a foot high, with a few broad roundish scales; flowers on short stalks, numerous, drooping, in a one-sided cluster, bluish, streaked with purple or dark red; upper lip of the corolla entire or slightly notched, lower lip divided into 3 lobes. In a copse at Lindridge near Kingsteignton, on roots of Laurel and Hazel. (E. B. t. 50.) P. IV. V.

Ord. LXII. SCROPHULARIACEÆ.

VERONICA. SPEEDWELL.

1. **V. serpyllifolia** (*thyme-leaved Speedwell.*)—Common in pastures and by waysides. Root fibrous; stems creeping, and throwing up numerous erect flowering branches; leaves in pairs, broadly oval, slightly crenate; flowers in a loose terminal spike, with leafy-looking bracts; corolla pale blue or white, with dark streaks; fruit inversely kidney-shaped. Torquay. Cockington. Shiphay, etc. (E. B. t. 1075.) P. V. VI.

2. **V. scutellata** (*marsh S.*)—In boggy places and by sides of ditches. Stems weak, slender, and spreading, 6 or 8 inches high; leaves opposite, sessile, linear and slightly toothed. Flowers in alternate loose lateral clusters, on slender stalks; corolla flesh-coloured, with purple streaks. Fruit flat and broad, deeply notched. Forde bog, near Newton. Bovey Heath. (E. B. t. 782.) P. VI.-VIII.

3. **V. Anagallis** (*water S.*)—In ditches and by edges of ponds. Plant varying in size from 6 inches to 2 feet; stems erect, succulent and branching; leaves sessile, opposite, lanceolate and serrated; flower-spikes opposite, rising from the axils of the leaves; flowers pale blue, with darker streaks; fruit small and roundish, notched at the top. Near Torre Abbey. Lanes about Marychurch. Forde bog. (E. B. t. 781.) P. VI.-VIII.

4. **V. Beccabunga** (*Brooklime.*)—In ditches and small brooks. Stems round and smooth, prostrate or floating, rooting at their joints, branched and leafy; leaves slightly stalked, oblong, somewhat toothed, succulent. Flowers small and blue, in axillary opposite clusters. Capsule roundish, slightly notched. Common. Ditches near Torre Abbey. Upton lanes. Ansti's Cove lane. Paignton lanes. (E. B. t. 655.) P. IV.-VIII.

5. **V. officinalis** (*common S.*)—In dry pastures and woods. Stems creeping, and rooting at the joints, from 6 inches to a foot

or more in length. Leaves opposite, oblong, serrated and hairy. Spikes lateral, and solitary from the axils of the leaves; flowers small, pale purplish-blue; capsule heart-shaped. Very abundant in pastures around Torquay, etc. (E. B. t. 765.) P. V.–VII.

6. **V. montana** (*mountain S.*)—In damp woods and hedges. Stem trailing, hairy; leaves stalked, ovate, coarsely toothed; clusters lateral, erect, few-flowered; flowers pale blue, with darker streaks; capsule very broad and flat, notched at the top and bottom. Near Torre Abbey. Copse by Kent's Cavern, etc. (E. B. t. 766.) P. IV.–VII.

7. **V. Chamædrys** (*Germander S.*)—In woods, fields, on hedgebanks, and by waysides. Stem trailing, from a few inches to a foot high, with a line of hairs on each side. Leaves opposite, sessile, ovate and pointed, sharply serrate; flower-stalks axillary, solitary, bearing beautiful clusters of large, brilliant blue flowers with rich purple veins; fruit small, inversely heart-shaped. Growing abundantly in all our lanes, etc. (E. B. t. 623.) P. IV.–VI.

8. **V. hederæfolia** (*ivy-leaved S.*)—In fields and waste places. Stems procumbent; leaves heart-shaped, but divided into from 5 to 7 unequal lobes, the end one being the largest, stalked; the 2 radical leaves oval, entire. Flower-stalks lateral, 1-flowered; corolla pale blue. Capsule of 2 swelling lobes. Very common under hedges and in tilled fields. (E. B. t. 784.) A. IV.–VIII.

9. **V. agrestis** (*green procumbent field S.*)—In fields and waste places. Stem procumbent; leaves all stalked, oval heart-shaped; flowers on short stalks, blue with the lower lip of the corolla white; capsule of 2 swelling lobes, hairy all over, with about 6 seeds in each cell. Very common, Warberry Hill. Walks at Meadfoot, etc. (E. B. S. t. 2603.) The *V. polita*, which is very similar, but with the petals wholly blue, is also found in the same localities. (E. B. t. 783.) A. IV.–IX.

10. **V. arvensis** (*wall S.*)—In sandy or gravelly fields and on walls. Stem upright, from 2 to 8 inches high, branched at the base; leaves opposite, ovate, hairy, serrate; upper leaves lanceolate, entire, having the appearance of bracteas, from the axils of which the flowers arise, forming altogether a loose terminal spike; corolla pale blue; capsule inversely heart-shaped, smooth. Very abundant in fields and on the top of nearly every wall. (E. B. t. 734.) A. IV.–VII.

BARTSIA. BARTSIA.

1. **B. viscosa** (*yellow viscid Bartsia.*)—In wet pastures.

Plant from 8 to 12 inches high, upright, not branched; leaves sessile, oblong or lanceolate, but very deeply lobed; root-leaves opposite, those of the stem alternate; whole plant downy and viscid. Flowers solitary, axillary, yellow; lower lip of the corolla striped with orange. Near the coal-pit at Bovey. Fields above the Castle at Dartmouth. (E. B. t. 1045.) A. VI.-X.

2. **B. Odontites** (*red B.*)—In cornfields, dry pastures, and waste places. Stem hairy, from 6 to 10 inches high, angular, branched. Leaves opposite, linear-lanceolate, distantly but sharply serrate. Flower-spikes numerous, crowning the stem; flowers purplish-red, with narrow leaf-like bracteas at the base of each; anthers yellow. Very common. (E. B. t. 1415.) A. VI.-VIII. *Euphrasia*, Bab.

EUPHRASIA. EYE-BRIGHT.

E. officinalis (*common Eye-bright.*)—In fields and pastures. A pretty little plant, seldom more than a few inches high; stem generally branched, but sometimes simple; leaves deeply toothed, ovate. Flowers axillary, but thickly crowded into a terminal head; upper lip of the corolla 2-lobed, under one divided into 3; flowers white or pinkish, with purple streaks. Very abundant in all our pastures. (E. B. t. 1416.) A. V.-IX.

RHINANTHUS. YELLOW-RATTLE.

R. Crista-galli (*common Yellow-rattle.*)—In meadows and pastures. Stem erect, from 1 to 2 feet high, quadrangular, branched, dotted with purple spots. Leaves sessile, opposite, lanceolate, acutely serrate. Flowers in the axils of the upper leaves, yellow; upper lip of the corolla beaked; calyx tumid, 4-cleft. Very frequent in our meadows, etc. (E. B. t. 657.) A. V.-VII.

MELAMPYRUM. COW-WHEAT.

M. pratense (*common yellow C.*)—In woods and thickets. Stem slender, from 12 to 18 inches; high branches opposite and branching. Leaves distant, in pairs, narrow-lanceolate, slightly heart-shaped at the base, or ovate. Flowers in pairs, axillary, large, of a pale yellow colour. Bradley wood, near

Newton. Berry Pomeroy woods. Holne Chase. (E. B. t. 113.) A. VI.–VIII.

PEDICULARIS. LOUSEWORT.

1. **P. palustris** (*marsh Lousewort.*) – In marshy and boggy places. Stem about a foot high, upright, branched; leaves pinnatifid, segments ovate, deeply lobed. Flowers pale crimson; calyx inflated, hairy. Capsule ovate, its oblique beak when ripe projecting beyond the calyx. Forde bog. Bovey heath. Moor near Chagford. (E. B. t. 399.) A. or P. ? V.–VII.

2. **P. sylvatica** (*pasture L.*)—Moist hilly pastures and heaths. Stem from 3 to 6 or 8 inches high, branched at the base, spreading; leaves pinnatifid, segments lobed. Flowers axillary, rose-colour, like the last in form, but larger; calyx not much inflated. Abundant, Warberry Hill. Hills beyond Meadfoot, etc. (E. B. t. 400.) A. or P. ? V.–VIII.

SCROPHULARIA. FIGWORT.

1. **S. nodosa** (*knotted Figwort.*)—In shady places, under hedges and in thickets. Root stout and tuberous, studded with small knobs. Stem from 2 to 3 feet high, square. Leaves ovate, acute, serrate, on short stalks. Flower-stalks axillary and terminal, forked; bracteas lanceolate. Flowers greenish, with the upper lip of the corolla dusky-purple; seed-vessel ovate and pointed. Ansti's Cove lane. Copse by Kent's Cavern, etc. (E. B. t. 1544.) P. VI. VIII.

2. **S. aquatica** (*water F.*)—In wet places and by sides of streams. Root fibrous, stem erect, square, scarcely branched, from 3 to 6 feet high; leaves oblong heart-shaped, opposite, coarsely serrate. Flowers in small, opposite, forked clusters, with bluntly linear bracts; corolla greenish, mixed with dark purple; capsule roundish. Frequent. Fields near Torre Abbey. Side of the Ansti's Cove lane. Wet meadows at Paignton, etc. (E. B. t. 854.) P. VI.–IX.

DIGITALIS. FOXGLOVE.

D. purpurea (*purple Foxglove.*)—Dry banks, waste places, under hedges, and in woods. One of the commonest, but the most stately and beautiful of our wild plants. Stem from 2 to 4 or 5 feet high, upright, leafy, terminated by its long flower-spike. Leaves stalked, alternate, ovate-lanceolate, serrate, downy beneath; the radical leaves very long. Flower-spike of very numerous bell-shaped flowers, of a rich purplish-crimson externally, lighter within, but sprinkled with purple spots. Very abundant. Meadfoot cliffs. Warberry Hill, etc. (E. B. t. 1297.) P. v.-viii.

ANTIRRHINUM. SNAPDRAGON.

1. **A. majus** (*great Snapdragon.*)—On old walls and clefts of rocks; very often but the outcast from gardens. Flowering stems arising from the leafy base of the plant from 1 to 2 feet high. Leaves narrow lance-shaped, entire; lower ones opposite upper alternate. Flowers large, in a crowded, handsome spike; corolla not spurred, more than an inch long, purplish-red and yellow; sometimes white. Walls near Torre Abbey. Babbicombe. Totness Castle. (E. B. t. 129.) P. vii. ix.

2. **A. Orontium** (*lesser S.*)—In dry sandy or gravelly fields. Stem erect, about 1 foot high, slender; leaves narrower than in the last, linear-lanceolate; flowers in loose spikes, distant, mostly in the axils of the upper leaves; corolla purple, small, surrounded by the long, linear segments of the calyx. Dartmouth. (E. B. t. 1155.) A. vii.-x.

LINARIA. TOADFLAX.

1. **L. Cymbalaria** (*ivy-leaved T.*)—On rocks, old walls, and shady banks. A trailing plant, with slender stems, rooting at the nodes; leaves 5-lobed, roundish heart-shaped, stalked, sometimes purplish on their under sides; flowers solitary, on long stalks; corolla small, pale lilac, with a yellowish palate; spur short. Very abundant on walls and rocks about Torquay. *Antirrhinum*, Linn. (E. B. t. 502.) P. iv.-x.

2. **L. spuria** (*round-leaved T.*)—Cornfields and waste places in a sandy soil. Stem trailing; leaves alternate, ovate, downy; flowers small, yellow; upper lip purple; corolla with a bent spur. Fields at Marychurch. (E. B. t. 691.) A. vii.-xi. *Antirrhinum*, Linn.

3. **L. Elatine** (*sharp-pointed Fluellin, or T.*)—In cornfields, etc., very similar to the last, but smaller in every part, and with broadly hastate leaves. Abundant in fields about Torquay. *Antirrhinum*, Linn. (E. B. t. 692.) A vii.-xi.

4. **L. repens** (*creeping T.*)—On rocky places near the sea

rare. Root creeping; stem from 12 to 18 inches high, slender and branched; lower leaves whorled, upper scattered, linear. Flowers in panicled clusters, lilac, with purple streaks, palate yellow. Near Kennock Farm, on the Downs west of Christowe, *Fl. D. Antirrhinum*, Linn. (E. B. t. 1253.) P. VII.-IX.

5. **L. vulgaris** (*yellow T.*)—In borders of fields and under hedges, frequent. Stem from 1 to 2 feet high, erect; leaves linear-lanceolate; flowers in a thickly crowded terminal spike large, yellow. Common under hedges and in waste places. Torquay. Marychurch, etc. *Antirrhinum Linaria*. (E. B. t. 658.) P. VII.-X.

6. **L. minor** (*least T.*)—In sandy fields. Stem from 4 to 10 inches high, branched; leaves generally alternate, linear-lanceolate, blunt; flowers solitary, on axillary flower-stalks, light purple, with yellow palate and white lower lip. Paignton common. Ilsington. Bovey Tracey. *Antirrhinum minus*. (E. B. t. 2014.) A. V.-X.

SIBTHORPIA. SIBTHORPIA, MONEYWORT.

S. Europæa (*creeping S., or Cornish Moneywort.*)—In shady boggy places, rare. A minute, but graceful plant, with creeping thread-like stems; leaves roundish kidney-shaped, alternate, gradually diminishing in size from the base to the ends of the stem. Flowers axillary, small and 5-cleft, the 3 upper segments rose-colour, the 2 under pale yellow. Boggy places by the side of the Dart, in Holne Chase. Between Newbridge and Spitchwick Lodge, *Fl. D.* (E. B. t. 649.) P. VII. VIII.

VERBASCUM. MULLEIN.

1. **V. Thapsus** (*great Mullein.*)—On banks, under hedges, and in waste ground. Stem upright, stout, from 3 to 5 feet high, woolly, angular. Leaves sessile, covered with white wool, ovate or oblong, decurrent at the base, and winging the stem. Flowers in a very dense terminal spike, rich yellow; stamens hairy, bearing red anthers. Very frequent. Meadfoot cliffs, etc. (E. B. t. 549.) B. VI.-VIII.

2. **V. Blattaria** (*moth M.*)—On banks and by borders of fields, not common. Stem from 3 to 5 feet high; leaves oblong, coarsely notched, lower ones stalked, upper ones sessile or clasping; flowers in a loose terminal cluster, bright yellow, stamens with purple hairs; the unopened buds tinged with orange. Was growing in 1856 on the cliffs above the New Road, and most

probably will be found there again. Berry Pomeroy. Dartmouth. Teignmouth. Chudleigh. (E. B. t. 393.) B. VI.–VIII.

3. **V. virgatum** (*large-flowered primrose-leaved M.*)—By roadsides and in fields. Stem from 2 or 3 to 6 feet high, very strong and stout, branching at the base. Root-leaves much like those of the Primrose, but lyrated and larger; stem-leaves oval-lanceolate, pointed, serrate, sessile, uppermost leaves clasping. Flowers in axillary clusters, from 2 to 6 in each axil, large, bright yellow, with purple-bearded stamens. Near the quarry at Meadfoot. Meadfoot cliffs. Lanes at Marychurch. Fields by Torre Abbey, etc. (E. B. t. 550.) B. VII. VIII.

4. **V. pulverulentum** (*yellow hoary Mullein.*)—By roadsides and in waste places. A bold, stately-looking plant, 3 or 4 feet high; stem and leaves covered with a mealy white wool, which may be easily rubbed off. Leaves broadly oblong and crenate, sessile, lowermost ones lengthened and narrowed into a stalk. Flowers numerous, in small clusters on the terminal panicle, yellow, white hairs to the filaments of the stamens. Has been found on Meadfoot cliffs, but not met with of late. (E. B. t. 487.) B. VII. VIII.

5. **V. Lychnitis** (*white M.*)—On banks, by waysides, and in fields, not frequent. Stem from 2 to 3 feet high, straight and angular; lower leaves stalked, oblong; upper leaves sessile, ovate, pointed, very woolly on their under sides; stem terminating in an ascending, branched panicle. Flowers numerous, small, very pale yellow. About Shillingford, near Exminster, *Fl. D.* (E. B. t. 58.) B. VII. VIII.

6. **V. nigrum** (*dark M.*)—On banks and by waysides. Called *nigrum* on the *lucus à non lucendo* principle, the plant itself being particularly bright and handsome. Stem upright, 2 or 3 feet high; lower leaves oblong heart-shaped, stalked; upper leaves nearly sessile, small, and pointed, both being crenate and slightly woolly beneath. Flower-spike terminal, elongated, with numerous clusters of flowers between the bracts, along its whole length. Flowers yellow and large, with purple clothed stamens. Meadfoot cliffs. Ilsham. (E. B. t. 59.) P. VI.–IX.

Ord. LXIII. **LABIATÆ.**

A. *Stamens* 2.

LYCOPUS. GIPSYWORT.

L. Europæus (*common Gipsywort.*)—On banks of rivers,

in ditches and marshes. An upright, branching plant; stems about 2 feet high, square, and rather rough. Leaves on short stalks, opposite, lanceolate, and very deeply toothed, almost pinnatifid. Flowers minute, in dense axillary whorls, whitish, with small purple dots. Goodrington Marsh. (E. B. t. 1105.) P. VI.–IX.

SALVIA. SAGE, CLARY.

S. Verbenaca (*wild English Clary, or Sage.*)—In dry pastures, and on rocky banks, etc., frequent. A coarse, upright, more or less hairy, slightly branched plant, from 1 to 2 feet high. Lower leaves stalked, ovate, very much wrinkled, lobed and toothed; upper leaves sessile and sharply serrate; floral bracts heart-shaped. Flowers in whorls of about six on the upright terminal spike; corolla purple. Park Hill. Teignmouth road. Daddyhole Plain, etc. (E. B. t. 154.) P. V.–VIII.

B. *Stamens* 4.

MENTHA. MINT.

1. **M. rotundifolia** (*round-leaved M.*)—In moist waste places. A coarse, green, hairy, and upright plant; leaves broadly oval, or roundish, whitish underneath, much wrinkled, sessile, sharply serrate. Flowers in whorls, on upright terminal spikes, pale pink, occasionally white. Fields near Torre Abbey. Blagdon. (E. B. t. 446.) P. VIII. IX.

2. **M. viridis** (*spear M.*)—In moist and marshy places. Stem erect; leaves acutely lanceolate, not hairy, serrated, without stalks; flower-spikes terminal and cylindrical. Flowers in whorls, rose-coloured. Near Exmouth, *Fl. D.* (E. B. t. 2424.) P. VIII.

3. **M. Piperita** (*Pepper M.*)—In watery places. Stem branched; leaves ovate-lanceolate, acute, sharply serrate, stalked; flower-spikes fuller and bearing larger whorls than the Spearmint. Flowers very pale purple. Plant from 1 to 3 feet high. Cockington. Chudleigh. (E. B. t. 687.) P. VIII. IX.

4. **M. aquatica** (*water M.*)—In marshes and wet ditches and on banks of rivers. Plant much branched, from 12 to 18 inches high, generally covered with soft hairs. Leaves stalked, ovate or slightly heart-shaped. Flowers in thick, terminal, round or oblong heads, with occasionally 2 or more additional whorls in the axils of the upper leaves, pink or purplish. *M. hirsuta,*

Linn. (E. B. t. 447.) Berry Pomeroy woods. Marshy fields at Paignton. The variety γ, *M. citrata*, grows in the Rocky Valley, near Torquay. (E. B. t. 1025.) P. VIII. IX.

5. **M. sativa** (*marsh whorled M.*)—On banks of streams and in wet situations. Apparently intermediate between *M. aquatica* and *M. arvensis;* more low and spreading than the former, and having its flowers in axillary whorls, without any spike or terminal head, and distinguished from the Corn Mint by its more tubular and longer calyx, as well as its larger flowers. This and its varieties may be found about Barton. Topsham. (E. B. t. 448.) *M. acutifolia*, E. B. t. 2415. *M. rubra*, Sm., E. B. t. 1413. P. VII.-IX.

6. **M. arvensis** (*corn M.*)—In cornfields and waste grounds. Root creeping; stems low, branched, and spreading, from 6 to 12 inches long; whole plant more or less hairy. Leaves stalked, ovate, and toothed; flowers in axillary whorls; calyx shallow, bell-shaped. Cornfields, etc., about Torquay, Cockington, and Marychurch. (E. B. t. 2119; and also var. β, *M. agrestis*, E. B. t. 2120; and γ, *M. gentilis*, E. B. t. 2118 and t. 449; *M. pratensis* of Hooker and Arnott.) P. VIII. IX.

7. **M. Pulegium** (*Penny-royal.*)—In marshy places and wet ditches. Smaller than the root of the Mints; stems prostrate; leaves small and ovate, downy with short hairs. Flowers in thickly crowded axillary whorls, pink; throat of the calyx hairy. Whole plant has a powerfully pungent smell. Forde bog, near Newton. (E. B. t. 1026.) P. VIII. IX.

THYMUS. THYME.

T. Serpyllum (*wild T.*)—On downs and dry pastures. Stems much branched, slender and procumbent. Leaves ovate or oblong, very small. Flowers generally six in a whorl, with exceedingly small floral leaves taking the place of bracts; whole plant hairy. Flowers numerous, purple. Park Hill. Daddyhole Plain. Babbicombe Down, etc. (E. B. t. 1514.) P. VI.-VIII.

ORIGANUM. MARJORAM.

O. vulgare (*common Marjoram.*)—On hedge-banks and dry bushy places. Stems from 8 or 10 inches to 2 feet high; leaves stalked, broadly ovate, slightly toothed; heads of flowers roundish, crowded, in a 3-branched terminal cluster. Flowers purple, with purple-tinged bracts. Very frequent. Hedge-banks on the Warberry Hill. Meadfoot cliffs, etc. (E. B. t. 1143.) P. VII.–IX.

TEUCRIUM. GERMANDER.

T. Scorodonia (*wood Germander.*)—In woods, and on dry, stony hedge-banks. Stem slightly branched, about 1 foot high, hairy; leaves stalked, ovate, heart-shaped at the base, much wrinkled, coarsely toothed, downy, both sides green. Flowers in lateral and terminal one-sided clusters, pale yellow, with prominent purplish-red stamens. Park Hill wood. Meadfoot cliffs. Copse, near Ansti's Cove, etc. (E. B. t. 1543.) P. VII. VIII.

AJUGA. BUGLE.

A. reptans (*common Bugle.*)—In woods and moist pastures. Flowering stems erect, rising from a tuft of stalked, obovate root-leaves, from 2 to 8 or 9 inches high, with nearly sessile ovate leaves, uppermost ones small and coloured. Flowers in crowded whorls in the axils of the leaves, blue. Very abundant everywhere about the neighbourhood. (E. B. t. 489.) P. V. VI.

BALLOTA. HOREHOUND.

B. nigra (*black Horehound.*)—Under hedges and in waste places. Stem from 2 to 3 feet high; leaves coarsely but sharply serrate; lower ones heart-shaped, upper ovate. Flowers in whorls, purple, sometimes, but rarely, white. Frequent by the roadsides, and borders of fields. *B. fœtida*, Bab. (E. B. t. 46.) P. VII. VIII.

LEONURUS. MOTHERWORT.

L. Cardiaca (*Motherwort.*)—Under hedges and in waste places, not common. Stem about 3 feet high, branched; lower leaves 5-cleft, broadly lobed and toothed; upper 3-lobed, entire, wedge-shaped at the base; uppermost lanceolate, undivided. Flowers in whorls, crowded, white with a purplish tinge. Near Canonteign. Chudleigh. North Bovey. Lustleigh. Teigngrace, *Fl. D.* (E. B. t. 286.) P. VIII.

H

GALEOPSIS. HEMP-NETTLE.

1. **G. Ladanum** (*red Hemp-Nettle.*)—In dry, waste places. Stem about 1 foot high, with opposite branches, covered with soft hairs, not swollen at the joints. Leaves on short stalks, lanceolate, slightly serrate, hairy on both sides. Flowers bluish-pink, with a mixture of crimson and white; calyx hairy, 5-cleft. Fields, etc., about Torquay and Marychurch. (E. B. t. 884.) A. VII.–IX.

2. **G. Tetrahit** (*common H.*)—In cornfields, woods, and cultivated grounds. Stem from 1 to 2 feet high, beset with sharp hairs, swollen below the joints. Leaves of an oblong, pointed, oval form, serrate, hairy. Flowers in whorls; corolla varies in colour from purple to white; calyx-teeth long and pointed. Very common. (E. B. t. 207.) A. VII.–IX.

GALEOBDOLON. WEASEL-SNOUT.

G. luteum (*yellow W., or Archangel.*)—In woods and damp shady situations. Stem erect, not branched, about 1 foot high; leaves bright green, ovate, pointed, sharply but unevenly serrate. Flowers in whorls, large and showy, yellow, with the under lip spotted with reddish-orange; upper lip of the corolla undivided and arching over; calyx-segments pointed and tipped with bristles. Ansti's Cove. Bradley Woods, in great abundance. (E. B. t. 787.) P. V. VI. [Placed by Bentham among the Lamiums, under the name of *Lamium Galeobdolon.*]

LAMIUM. DEAD-NETTLE.

1. **L. album** (*white Dead-Nettle.*)—Under hedges, in borders of fields and waste places. Stem from 12 to 18 inches high; leaves in pairs, stalked, heart-shaped, pointed, coarsely toothed; flowers in close axillary whorls, pure white; upper lip undivided, arched; calyx-teeth long, sharp, and spreading. Chelston. Cockington lanes, etc., frequent. (E. B. t. 768.) P. IV.–IX.

2. **L. purpureum** (*red D.*)—Under hedges, in fields and waste places. Stem from 4 to 8 or 9 inches high, with a few roundish stalked leaves at the base, then rising naked, but thickly covered at the top with broadly heart-shaped, unevenly toothed leaves on short stalks. Flowers purplish-red; floral leaves with a purplish tinge. Very abundant. (E. B. t. 769.) A. IV.–X. *I have found this plant in flower during the whole year, in sheltered spots about Torquay.*

BETONICA. BETONY.

B. officinalis (*wood Betony.*)—Under hedges and in woods and thickets. Stem from 1 to 2 feet high, erect and simple; root-leaves on long stalks, ovate, serrate; upper leaves nearly sessile, oblong, toothed; the pairs of leaves distant from each other; whorls of flowers forming a dense, oblong terminal spike; corolla of dull purplish rose-colour, upper lip flattish. Very common. *Stachys Betonica,* Bab. and Benth. (E. B. t. 1142.) P. VI.–VIII.

STACHYS. WOUNDWORT.

1. **S. sylvatica** (*hedge Woundwort.*)—In woods and under hedges. Stem from 2 to 4 feet high, stout, erect, slightly branched. Leaves large, ovate, cordate, strongly serrate. Flower-spike terminal, composed of numerous close whorls of from 6 to 10 flowers in each; corolla of a deep purplish-red, upper lip lip vaulted. Frequent. Walks at Meadfoot. Copse near Ansti's Cove. Cockington and Shiphay lanes. (E. B. t. 416.) P. VII. VIII.

2. **S. palustris** (*marsh W.*)—In ditches, on river-banks, and in wet places. Root creeping extensively. Stems 2 or 3 feet high, stout; lower leaves slightly stalked; upper ones half clasping, oblong or lanceolate, serrated, rather woolly beneath. Spike of many whorls, with a pair of leaves beneath each whorl; corolla purplish-red. Common. Osier-beds at Paignton, etc. (E. B. t. 1675.) P. VII. VIII. Var. β. of Hooker and Arnott, Marychurch.

3. **S. Germanica** (*downy W.*)—In fields and under hedges. Plant from 1 to 3 feet high; stem as well as leaves covered with whitish silky hairs, giving the whole plant a hoary appearance. Leaves ovate, pointed, serrate, the radical ones on long stalks, the others almost sessile, all very much veined. Flowers numerous, in axillary whorls; corolla pinkish, streaked with white, upper lip arched. Fields on the Warberry Hill, *Mr. Earle.* (E. B. t. 829.) P. VII.

4. **S. arvensis** (*corn W.*)—In cornfields. Stem weak, prostrate or ascending, branched. Leaves opposite, on short stalks, ovate or slightly heart-shaped; floral leaves oval-oblong, pointed, sessile. Flowers in whorls of from 4 to 6, small; corolla pale purple, upper lip vaulted. Frequent in tillage fields. Walks on Meadfoot cliffs. Warberry. (E. B. t. 1154.) A. VIII. IX.

NEPETA. CAT-MINT, GROUND-IVY.

1. **N. Cataria** (*Cat-mint.*)—In fields, hedges, and waste places. Stems from 2 to 3 feet high, erect, downy; leaves stalked, heart-shaped, serrate, downy beneath; floral leaves bract-like. Flowers in spiked, shortly stalked whorls, white, with a pinkish tinge, and spotted with rose-colour. Babbicombe road. Cockington. Shiphay lanes, etc. (E. B. t. 137.) P. VII.–IX.

2. **N. Glechoma** (*ground Ivy.*)—In thickets, hedges, and waste grounds. Stem extensively creeping; leaves stalked, kidney-shaped, crenated, downy. Whorls on one side of the stem, axillary, 3- to 4-flowered; flowers large and blue. Very abundant. *Glechoma hederacea*, Linn. (E. B. t. 853.) P. III.–V.

MARRUBIUM. WHITE HOREHOUND.

M. vulgare (*common white Horehound.*)—By roadsides, borders of fields, and waste places. Stem thick and cottony, branched; leaves stalked, roundish, irregularly serrate, wrinkled and hoary. Plant from 12 to 18 inches high. Flowers in crowded whorls in the axils of the upper leaves, small, of a dirty-white colour; upper lip of the corolla 2-cleft; calyx with from 5 to 10 small hooked teeth. Fields on the Warberry Hill. Watcombe. Waste spots at Marychurch. (E. B. t. 410.) P. VIII. IX.

CALAMINTHA. CALAMINTH, BASIL-THYME, WILD BASIL.

1. **C. Acinos** (*common Basil-Thyme.*)—In cultivated fields and waste places. Plant 6 or 8 inches high, branched, slightly downy. Leaves on short stalks, small, oblong, slightly toothed, but at times almost entire. Flowers in axillary whorls, on short upright stalks; corolla pale-purple or white. Frequent in tillage fields and dry pastures. Daddyhole Plain, on the slope near the wall of Rock End grounds. Meadfoot cliffs. Fields near Hope's Nose. *Thymus*, Linn. (E. B. t. 411.) A. VII. VIII.

2. **C. officinalis** (*common Calaminth.*)—In woods, borders of fields, and by roadsides. Stem from 1 to 2 feet high, strong and erect, with straggling branches, hairy. Leaves stalked, broadly ovate, toothed. Flowers in loose lateral cymes, lilac; lower lip of the corolla dotted. Roadside near Chudleigh, Ilsington, *Fl. D. Thymus Calamintha.* (E. B. t. 1676.) P. VII.–IX.

3. **C. Clinopodium** (*common wild Basil.*)—In woods and under hedges. Stem from 12 to 18 inches high, covered with soft hairs, rather wavy. Leaves stalked, ovate, slightly serrate, hairy; whorls both terminal and axillary; flowers on hairy stalks, surrounded by linear, hairy floral leaves or bracts; corolla large, purple. Frequent. On the cliffs near Ilsham beach. *Clinopodium vulgare*, Linn. (E. B. t. 1401.) P. VII.–IX.

MELITTIS. BASTARD-BALM.

M. Melissophyllum (*Bastard-Balm.*)—In woods, thickets, and shady places. A very handsome plant, and worthy of a place in our gardens. Stem from 1 to 2 feet high; leaves large, stalked, heart-shaped, and coarsely serrate. Flowers large, in axillary whorls; pink, variegated with white, and variously spotted with purple; corolla-tube broad, nearly an inch long; the upper lip concave and thrown back, the lower one 3-lobed and spreading. Park Hill wood. Copse near Ansti's Cove. (E. B. t. 577.) P. V. VI. *The M. grandiflora is merely a variety of this, and not a distinct species.*

PRUNELLA. SELF-HEAL.

P. vulgaris (*common Self-heal.*)—In moist pastures, fields, and hedges. A low branching plant; leaves stalked, ovate or oblong, slightly waved but not toothed. Flowers in closely packed terminal whorls, forming a dense spike, with broad, bract-like floral leaves beneath each whorl; corolla violet-blue, sometimes reddish. Common in fields, etc., about Torquay and Marychurch. (E. B. t. 961.) P. VII. VIII.

SCUTELLARIA. SKULL-CAP.

1. **S. galericulata** (*common Skull-cap.*)—In damp, shady, stony places and banks of streams. A weak-looking, slightly downy plant; stem erect, from 8 to 12 inches high, somewhat branched; leaves opposite, nearly sessile, lanceolate but heart-shaped at the base, crenate. Flowers axillary, solitary or in pairs, nearly sessile, drooping slightly; corolla blue, lower lip with white streaks. Berry Pomeroy woods. Bovey Heath. Exminster Marsh. Fingle Bridge, near Chagford. (E. B. t. 523.) P. VII. VIII.

2. **S. minor** (*lesser S.*)—In bogs and moist heathy places. Smaller and more slender than the last, from 4 to 6 inches high. Leaves entire; lower ones broadly ovate and somewhat lobed at the base. Flowers small, on short axillary stalks, rose-colour; the lower lip white, with pink spots. Forde bog, near Newton. Wet places about Dartmoor. (E. B. t. 524.) P. VII.-X.

Ord. LXIV. VERBENACEÆ.

VERBENA. VERVAIN.

V. officinalis (*common Vervain.*)—By roadsides and in dry pastures. Stem erect, from 1 to 2 feet high, rather hairy; leaves opposite, lanceolate but deeply cut, almost pinnatifid, upper ones 3-lobed; stem dividing at the top into 5 or 6 slender flower-spikes; flowers small and sessile, each accompanied by a small, ovate, pointed bract; corolla pale purple, tubular, 5-cleft. Frequent in dry waste places. Warberry Hill. Babbicombe, etc. (E. B. t. 767.) P. VII.-IX.

Ord. LXV. LENTIBULARIACEÆ.

PINGUICULA. BUTTERWORT.

1. **P. vulgaris** (*common Butterwort.*)—In bogs and moist heaths. Leaves all radical, light green, ovate, covered with small crystalline points, the margins rolled up. Flower-stalks rising naked from the tuft of leaves, each bearing a single bluish-purple flower, with a long spur to the corolla. Forde bog, near Newton. Bovey Heath. (E. B. t. 70.) P. V.-VII.

2. **P. Lusitanica** (*pale B.*)—In similar situations to the last. Like the *P. vulgaris*, but smaller; leaves more delicate, and covered with red veins, margins very much curled up; flower-stalks slender; flower pale yellow tinged with lilac, spur short but much curved, lobes of the corolla nearly equal. Forde bog. Haldon. Bogs in Dartmoor, Ivybridge, etc. (E. B. t. 145.) P. VI.-X.

UTRICULARIA. BLADDERWORT.

U. vulgaris (*greater Bladderwort.*)—In ditches and pools.

The floating branches, extending from 6 inches to a foot, bear numerous hair-like much divided leaves, to which are attached a multitude of little curved air-bladders; the flower-stalk rises upright a few inches out of the water, and bears a loose cluster of handsome alternate yellow flowers, the flower-stalks and calyx being tinged purple. Stover canal-head. Bovey Heath. Ponds between Teignbridge and Kingsteignton, Powderham marshes, *Fl. D.* (E. B. t. 253.) P. VI. VII. *U. intermedia*, with some of the branches bearing leaves without vesicles, is considered by Bentham to be merely a barren form of *U. minor*. The 'Flora Devoniensis' gives as its habitat a pool between Teignbridge and Kingsteignton. (E. B. t. 2489 and 254.)

Ord. LXVI. **PRIMULACEÆ**.

PRIMULA. PRIMROSE, OXLIP, COWSLIP.

1. **P. vulgaris** (*common Primrose.*)—In woods, pastures, and on hedge-banks. Leaves more or less hairy, ovate or oblong, large, slightly toothed and much wrinkled, of a pale fresh green; flower-stalks, arising apparently single, but really from a common stalk so short as to be concealed by the leaves, bear each a terminal, pale yellow flower, with a deeply 5-lobed spreading corolla. Abundant in all our lanes and hedges. (E. B. t. 4.) P. IV. V.

2. **P. veris** (*common Cowslip.*)—In meadows and pastures. Leaves like the last but not hairy; flower-stalks bearing an umbel of flowers; corolla deeper yellow and much smaller than that of the Primrose, its segments not so widely spread. Very frequent. Warberry Hill. Cockington lanes. Shiphay, Ansti's Cove, etc. (E. B. t. 5.) P. IV. V.

3. **P. elatior** (*Oxlip.*)—Similar situations to the last. Leaves ovate, toothed and contracted below. Flower-stalk bearing a many-flowered umbel; segments of the corolla more dilated than those of the Cowslip, and having more the appearance of the Primrose; outer flowers of the umbel usually drooping. Not so frequent as the last. New walks at Ansti's Cove. (E. B. t. 513.) P. IV. V.

GLAUX. SEA MILKWORT.

G. maritima (*sea M., or black Saltwort.*)—On seashores and in salt-marshes. Plant from 2 to 6 inches high; stem branched, stout and fleshy; leaves opposite, sessile, ovate, very small and succulent. Flowers rose-coloured, sessile and axillary, with a 5-lobed calyx and destitute of a corolla. Paignton sands. Goodrington sands. Exminster Marsh. (Hackney, near Kingsteignton, *Fl. D.*) (E. B. t. 13.) P. VI. VII.

LYSIMACHIA. LOOSESTRIFE.

1. **L. vulgaris** (*great yellow Loosestrife.*)—By sides of rivers and in wet shady places. Plant from 2 to 4 feet high, with straight leafy stems. Leaves usually opposite, but sometimes growing 3 or 4 in a whorl, ovate-lanceolate, on very short stalks. Flowers in lateral and terminal clusters, large and handsome, yellow. Bank of canal at Teigngrace. Side of Stover canal. Islet in the Dart above Staverton Bridge. (E. B. t. 761.) P. VII. VIII.

2. **L. nemorum** (*yellow Pimpernel, or wood L.*)—In woods and shady places. Stems procumbent and rooting from the lower joints, reddish; leaves ovate, pointed, opposite and stalked. Flowers on long axillary stalks, solitary; corolla 5-lobed, spreading, bright yellow. Bradley woods, near Newton. Berry Pomeroy woods. Gidleigh near Chagford. Whyddon Park, etc. (E. B. t. 527.) P. V.–VIII.

ANAGALLIS. PIMPERNEL.

1. **A. arvensis** (*scarlet P., or Poor Man's Weather-glass.*)—In cornfields and by waysides. Stems procumbent or ascending, branched; leaves opposite, sessile, ovate. Flowers on long axillary flower-stalks, bright scarlet; corolla opening out wide in bright sunshine, and closing up in cloudy weather. Abundant everywhere. (E. B. t. 529.) *A. cærulea* Professor Henslow has ascertained to be a variety of this. (E. B. t. 1823.) A. VI. VII.

2. **A. tenella** (*bog Pimpernel.*)—In bogs and on wet mossy banks. An exceedingly delicate and pretty little plant; stems from 2 to 4 inches long, branched and creeping; leaves opposite, roundish-ovate; flowers on long slender footstalks, large, of a most lovely rose-colour. Sometimes in the turf of the public garden. Forde bog, near Newton. Holne Chase. Ivybridge. Boggy places about Chagford, etc. (E. B. t. 530.) P. VII. VIII.

CENTUNCULUS. CHAFFWEED.

C. minimus (*small C., or bastard Pimpernel.*)—In moist sandy or gravelly places. A very diminutive plant, from 1 to 2 inches high; stem slightly branched at the base; leaves small, ovate. Flowers solitary, sessile, very small, pale pink; corolla 4-cleft. Petit Tor, near Marychurch. Bovey Heath. (E. B. t. 531.) A. VI. VII.

SAMOLUS. BROOKWEED.

S. Valerandi (*Brookweed, or water Pimpernel.*)—In marshy and watery places. Plant from 6 to 10 inches high; leaves alternate, roundish-oval, on short stalks; flowers in terminal clusters, small and white; corolla with 5 spreading segments. Petit Tor, by a little streamlet near the beach. Formerly at Meadfoot. (E. B. t. 703.) P. VII. VIII.

ORD. LXVII. PLUMBAGINACEÆ.

ARMERIA. THRIFT, SEA-PINK.

A. maritima (*common T., or S., or Sea-Gilliflower.*)—On muddy or rocky seashores. Root-stock throwing up a dense tuft of narrow-linear leaves. Heads of flowers on long leafless stalks; flowers closely crowded together, pink, sometimes white. Shores of Torbay, Meadfoot, etc. *Statice Armeria.* (E. B. t. 226.) P. IV.–IX.

STATICE. SEA-LAVENDER.

1. **S. Limonium** (*spreading-spiked Sea-Lavender.*)—In salt-marshes and on rocks by the sea. Leaves all radical, from 2 to 6 inches long, ovate-lanceolate, narrowing at the base into a stalk. Flower-stalk, upright, leafless, repeatedly divided at the top, and forming a thick panicle of bluish-purple flowers. Exminster marshes. (E. B. t. 102.) B. VII.–IX.

2. **S. binervosa** (*upright-spiked S.*)—On rocks near the sea. Leaves much smaller than the last, and narrowed into a winged stalk at the base; flower-stalk branched from below the middle; spikes compact and erect; flowers large, of a purplish blue. Cliffs

above Thunder Hole. Paignton cliffs. Berry Head. (*S. spathulata*, E. B. S. t. 2663.) *S. occidentalis*, Bab. *S. auriculæfolia*, Benth. B. VII. VIII.

ORD. LXVIII. PLANTAGINACEÆ.

PLANTAGO. PLANTAIN.

1. **P. major** (*greater Plantain.*)—By waysides, in pastures and waste places. Leaves radical, broadly ovate, on long coarsely ribbed stalks, more or less toothed or waved. Flower-stalks many, upright, surmounted by a long, pointed spike of numerous closely crowded flowers; corolla white; flowers at the base of the spike opening first; seed-vessel ovate, with from 4 to 8 seeds in each cell. Very abundant. (E. B. t. 1558.) P. VI.–VIII.

2. **P. media** (*hoary P.*)—In meadows and pastures. Leaves growing from the root-stock, pressing close to the ground, sessile, ovate, downy. Flower-stalks tall, pubescent; spikes cylindrical; stamens long, with yellow anthers, and shining purple filaments; capsule with 1 seed in each cell. Barton, *Mr. C. E. Parker.* (E. B. t. 1559.) P. VI.–IX.

3. **P. lanceolata** (*Ribwort P.*)—In meadows and pastures. Leaves all radical, lanceolate, ascending, 5-ribbed, tapering at both ends; spikes ovate or cylindrical, on long stalks; flowers with a blackish ovate bract at the base of each, stamens yellow; 1 seed in each cell of the capsule. Common everywhere. (E. B. t. 507.) P. VI. VII.

4. **P. maritima** (*seaside P.*)—Pastures and marshy places by the seaside. Leaves numerous, linear, fleshy, usually entire but sometimes toothed. Spike cylindrical, on a round flower-stalk; flowers numerous; cells of capsule 1-seeded. A very variable plant in size and appearance. Paignton, Goodrington, etc. (E. B. t. 175.) P. VI.–IX.

5. **P. Coronopus** (*buck's-horn P.*)—On the seacoast, and in barren gravelly soils inland. Leaves all springing from the tapering root, spreading, pinnately divided into numerous linear segments, lying closely upon the ground. Flower-stalk round, hairy; spike slender, compact; flowers white; capsule of four 1-seeded cells. Meadfoot cliffs. Paignton sands. Teignmouth sands. Bovey Heath. (E. B. t. 892.) A. VI. VII.

LITTORELLA. SHOREWEED.

L. lacustris (*Plantain Shoreweed.*)—In watery places and margins of lakes. Leaves all radical, linear and fleshy, slightly channelled. Flowers white, male flowers on long stalks, female flowers sessile among the leaves, and without a calyx. Bovey Heath. Haldon. (E. B. t. 468.) P. VI. VII.

Sub-Class IV. MONOCHLAMYDEÆ.
(Ord. LXIX.–LXXXVII.)

Ord. LXIX. AMARANTHACEÆ.

Ord. LXX. CHENOPODIACEÆ.

Subord. I. *CYCLOLOBEÆ.*

Tribe I. Chenopodieæ.

BETA. BEET.

B. vulgaris (*common Beet.*)—On the seashore and waste places near the sea. Root thick and fleshy; stems numerous, procumbent; leaves succulent, triangularly ovate, narrowed at the base into a footstalk; the upper stem-leaves oblong, sessile, all with wavy edges; flower-spikes long, simple, and leafy; flowers in twos or threes, greenish-yellow; calyx 5-partite; no corolla. *B. maritima*, Linn. Babington considers it should still retain the last name as being a distinct species from the common Beet. Common on our shores. Meadfoot. Paignton sands. Goodrington. Teignmouth, etc. (E. B. t. 285.) P. VI.–IX.

CHENOPODIUM. GOOSEFOOT.

1. **C. olidum** (*stinking Goosefoot.*)—Under walls and in waste places. Stems numerous, spreading, branched, and leafy; leaves stalked, alternate, angularly ovate, covered, as is the whole plant, with a greasy fetid down. Flowers in thickly clustered spikes, greenish; corolla, as in all the species of this genus, absent. Teignmouth. Chudleigh. (E. B. t. 1034.) A. VIII. IX.

2. **C. polyspermum** (*many-seeded G.*)—In cultivated and waste places. Stems prostrate and spreading; leaves alternate, stalked, ovate, blunt, dark green, and free from any mealiness. Clusters of flowers small, in axillary spikes; flowers green. Seed

black, partially covered by the 5-cleft calyx. *C. acutifolium* (E. B. t. 1481), with upright stem, and ovate, pointed leaves, is pronounced both by Hooker and Arnott, and Babington, to be undistinguishable from this. Waste places near Teignmouth. (E. B. t. 1480.) A. VIII. IX.

3. **C. murale** (*nettle-leaved G.*)—Beneath walls, by roadsides, and in waste rubbishy places. Plant sometimes erect, sometimes decumbent, a foot or more high; leaves ovate, acute, and sharply serrate; flowers in compound clusters from the sides and summit of the stem; calyx almost closing over the seed. Very frequent. Marychurch. Teignmouth, etc. (E. B. t. 1722.) A. VIII. IX.

4. **C. album** (*white G.*)—In waste places and about dungheaps. The most common of the genus. Stem branched, angular, from 1 to 2 feet high; leaves ovate, slightly angular, irregularly jagged; uppermost leaves oblong and entire. Flowers clustered in short axillary spikes, with a terminal spike crowning the stem; calyx quite investing the seed. Very common. (E. B. t. 1723.) A. VII.-IX.

5. **C. rubrum** (*red G.*)—On dunghills and under walls. Stems 1 to 2 feet high, upright, branched, furrowed; leaves alternate, of a triangular form, irregularly toothed. Clusters of flowers in upright axillary spikes. Paignton. Teignmouth, etc. (E. B. t. 1721.) A. VIII. IX.

6. **C. Bonus-Henricus** (*Mercury G., or Good-King-Henry.*)—Stem about 1 foot high, slightly branched; leaves large, broadly triangular, resembling spinach-leaves, alternate and stalked. Flowers in lateral and terminal, clustered and leafless spikes, green. The young plants are frequently eaten instead of spinach. Paignton. Kingskerswell. Chudleigh. (E. B. t. 1033.) P. V.-VIII.

TRIBE II. ATRIPLICEÆ.

ATRIPLEX. ORACHE.

1. **A. portulacoides** (*shrubby Orache, or Sea-Purslane.*)—On the seashore. A straggling, much branched shrub, from 12 to 18 inches or 2 feet high, covered with grey, silvery scaliness; lower leaves oblong or obovate; upper leaves lanceolate or linear as they ascend the stem. Flowers in short axillary spikes, and in a terminal panicle, small, dull yellow. Sterile and fertile flowers on the same plant in all this genus. Budleigh Salterton, *Miss A. Griffiths.* (E. B. t. 261.) P. VIII.-X. *Obione*, Bab.

2. **A. laciniata** (*frosted sea O.*)—On sandy seashores. Plant covered with a white scaly meal; stem procumbent and

spreading, rose-coloured; leaves broadly triangular and coarsely toothed. Male flowers in dense spikes; female sessile and axillary. Teignmouth sands. Exmouth. *A. arenaria*, Bab.; *A. rosea*, Benth. (E. B. t. 165.) A. VII.-IX.

3. **A. patula** (*spreading halberd-leaved O.*)—In salt-marshes and by the seaside. Plant erect or procumbent; leaves stalked; the lower ones broadly triangular, coarsely and irregularly toothed; uppermost lanceolate and entire. Flowers clustered in slender spikes, in narrow, leafy, terminal panicles. Paignton sands. Goodrington sands. E. B. t. 936; *A. hastata*, Bab.; *A. angustifolia*, E. B. t. 1774, with lanceolate lower leaves and linear upper ones. Paignton. *A. deltoidea*, E. B. S. t. 2860, with lower leaves halberd-shaped and unequally toothed. Meadfoot, Mr. C. E. Parker. These two last Bentham classes as varying forms of *A. patula.* A. VI.-X.

TRIBE III. SALICORNEÆ.

SALICORNIA. GLASSWORT.

S. herbacea (*jointed Glasswort.*)—On the seashore and muddy salt-marshes. A leafless plant, with a jointed herbaceous stem, erect and branched, from 3 to 12 inches high; joints compressed, thickening upwards and hollowed out on each side. Flowers sessile, 3 at the base of each joint of the terminal spikes. Goodrington Marsh. (E. B. t. 415. With the stem procumbent, *S. procumbens*, E. B. t. 2475.) A. VIII. IX.

SUBORD. II. *SPIROLOBEÆ.*

TRIBE IV. SUÆDEÆ.

SUÆDA. SEA-BLITE.

1. **S. fruticosa** (*shrubby Sea-Blite.*)—On the seacoast, rare. Stem from 2 to 3 feet high, shrubby, with numerous upright leafy branches; leaves succulent, blunt, and semi-cylindrical. Flowers very small, solitary and sessile, green. Budleigh Salterton, *Miss A. Griffiths.* (E. B. t. 635.) P. VII. VIII.

2. **S. maritima** (*annual S.*)—On the seashore, frequent. A much smaller plant than the last. Stem erect or procumbent, herbaceous, much branched and leafy; leaves semi-cylindrical, pointed, smooth, and succulent. Flowers sessile, green, with a

pair of narrow ovate bracteas to each. Starcross. Exmouth Warren. Budleigh Salterton. (E. B. t. 633.) A. VII.-IX.

TRIBE V. SODEÆ.

SALSOLA. SALTWORT.

S. Kali (*prickly S.*)—On sandy seashores. Stem much branched and extensively spreading, stiff and angular, from 6 inches to a foot long; all the leaves terminated by a stout sharp thorn. Flowers in the upper axils, sessile, pale-greenish, solitary. Paignton and Goodrington sands. (E. B. t. 634.) A. VIII.

ORD. LXXI. SCLERANTHACEÆ.

SCLERANTHUS. KNAWEL.

S. annuus (*annual Knawel.*)—Frequent in cornfields. Plant with numerous spreading stems, much branched; leaves linear, with a membranous margin at their base. Flowers green, solitary in the axils of the stem or in terminal leafy clusters. Very common. (E. B. t. 351.) A. VI.-VIII.

ORD. LXXII. POLYGONACEÆ.

POLYGONUM. PERSICARIA, BISTORT, KNOT-GRASS, BUCKWHEAT.

1. **P. aviculare** (*common Knotgrass.*)—In waste places and cultivated ground. A tough, much branched plant, from a few inches to 2 feet long; stem erect or procumbent; leaves oblong, narrow, alternate. Flowers axillary, on short stalks, in clusters of from 2 to 5, small, variegated with red, white, and green; nut triangular. Abundant. (E. B. t. 1252.) A. V.-IX.

2. **P. Roberti** (*Robert's K.*)—Sandy seashores in the west of England. Stem long and straggling, decumbent; leaves distant, elliptic-lanceolate, succulent, curled at the margin. Flowers small; nut protruding, smooth and shining. Paignton sands. *P. Raii*, Bab. (E. B. S. t. 2805.) A. VII.-IX.

3. **P. Fagopyrum** (*common Buckwheat.*)—About dunghills and in cultivated places. Stem erect, about 1 foot high, wavy;

leaves arrow-shaped, cordate at the base. Flowers in spreading terminal clusters, red and white mingled. Not indigenous, sown as food for game and poultry. In fields and about farmyards. *Fagopyrum esculentum*, Bab. (E. B. t. 1044.) A. VII. VIII.

4. **P. Convolvulus** (*climbing Buckwheat.*)—In fields and hedges. Stem from 1 or 2 to 5 or 6 feet long, prostrate or climbing, and twisting around the stems and branches of neighbouring plants. Leaves stalked, alternate, heart-shaped, and tapering to a point at their ends, wavy. Flowers in short spikes, rising from the axils of the upper leaves, greenish-white with a pink tinge. Too common. (E. B. t. 941.) A. VII.-IX.

5. **P. amphibium** (*amphibious Persicaria.*) — In ponds, ditches, and damp ground. Stem from 2 to 3 feet long, round, very little branched when growing in water, clothed with tubular membranous sheaths or stipules, out of which the leaves appear to arise. Leaves ovate-oblong or narrow-lanceolate, or varying between these forms, according to situation. Flower-spikes erect, on a reddish flower-stalk; flowers of a bright rose-colour. Goodrington. (E. B. t. 436.) P. VII.-IX.

6. **P. Persicaria** (*spotted Persicaria.*)—In ditches and damp waste ground. Stem erect and spreading, from 1 to 2 feet high, branched, sometimes slightly pubescent, reddish. Leaves on short stalks, upper ones nearly sessile, lanceolate, marked in the centre with a dark blotch. Flower-spikes numerous, terminal, on long stalks. Flowers rose-coloured or whitish-green. Stipules or sheaths fringed at the top with short, fine bristles. Meadows near the sea at Paignton. (E. B. t. 756.) A. VI.-X.

7. **P. lapathifolium** (*pale-flowered Persicaria.*)—In fields, by roadsides, and on dunghills. Bentham says that this is probably a mere variety of *P. Persicaria*. It is very variable in its appearance; it is distinguished by having its pedicels and perianths dotted with small, prominent glands, by its stipules not being fringed, and by its doubly concave nuts. The flowers vary from whitish to pale green and red. Paignton. Waste piece of ground near Forde bog. (E. B. t. 1382.) A. VII. VIII.

8. **P. Hydropiper** (*biting Persicaria.*) — In ditches and watery places. Stem from 1 to 3 feet high, erect; leaves lanceolate, waved. Stipules usually with scattered glands. The slender, drooping spikes distinguish it from all the other species. Flowers distant, reddish. Whole plant pungent and acrid to the taste. Forde bog. Wet meadows at Paignton. (E. B. t. 989.) A. VIII. IX.

RUMEX. DOCK, SORREL.

1. R. Hydrolapathum (*great Water-Dock.*)—In ditches and on the sides of rivers and pools. Root large and knobby; stem from 3 to 5 feet high; leaves lanceolate; lower leaves very large, sometimes as much as 18 inches long, heart-shaped at the base. Flowers in crowded whorls, on long terminal spikes. River Exe, near Countess Wear, *Fl. D. R. aquaticus*, Sm. (E. B. t. 2104.) P. VII. VIII.

2. R. crispus (*curled D.*)—By roadsides, in pastures, and waste places. Plant 2 or 3 feet high; stem branched; leaves stalked, lanceolate, acute, very much waved. Flower-whorls very numerous, much crowded when in fruit; valves or enlarged sepals cordate. Very common in our fields and pastures. (E. B. t. 1998.) P. VI.-VIII.

3. R. obtusifolius (*broad-leaved D.*)—In waste places and by waysides. Stem 2 or 3 feet high, not much branched. Root-leaves very large, obtusely heart-shaped, stalked, crisped at the margin; upper leaves oblong or lanceolate. Flower-whorls distant; segments of the fruit-covering with 3 small teeth near the base. Common everywhere. (E. B. t. 1999.) P. VII.-IX.

4. R. sanguineus (*bloody-veined D.*)—Shady pastures and woody places, rare. Stem 2 or 3 feet high; leaves ovate-lanceolate, stalked. Lower leaves large, and either heart-shaped or rounded at the base. Whorls distant, on long, alternate, leafless branches. Veins of the leaves bright red. In an orchard near Beckey Fall, *Fl. D.* (E. B. t. 1533.) The variety β of Hooker and Arnott. *R. viridis*, with green veins to the leaves, frequent. Cockington lanes. Lane between the Warberry Hill and Babbicombe, etc. P. VI.-VIII.

5. R. conglomeratus (*sharp D.*)—In marshy meadows and wet places. Very much like, but distinguished from *R. viridis* by its leafy flower-stalks, nearly every whorl being accompanied by a leaf. Berry Pomeroy woods. *R. acutus*, Sm. (E. B. t. 724.) P. V.-VIII.

6. R. pulcher (*fiddle D.*)—By roadsides and in dry waste places. Stem procumbent and spreading, with straggling branches; lower leaves oblong, stalked, cordate at the base, and narrowed in below their centre, making them somewhat fiddle-shaped. Whorls distant and leafy; flowers in close clusters. Frequent. Torquay, Cockington, etc. (E. B. t. 1576.) P. V.-VIII.

7. R. maritimus (*golden D.*)—In marshes near the sea, but sometimes found in inland situations. Stem from 12 to 18 inches high, branched and leafy; leaves lanceolate, narrow, not waved. Flowers very small, in numerous dense leafy whorls. When in seed the whole plant is frequently of a golden-yellow colour. Banks of the canal near Exeter, *Fl. D.* (E. B. t. 735.) P. VII. VIII.

8. **R. palustris** (*yellow marsh D.*)—In marshy and boggy situations. Very much like the last, with linear-lanceolate leaves, but with the flower-whorls much more distant from each other, and the flowers not so crowded. Bovey Heathfield, *Fl. D.* (E. B. t. 1932.) P. VII.-IX.

9. **R. Acetosa** (*common Sorrel.*)—In meadows and pastures, very common. Stem from 1 to 2 feet high, rising generally from a loose cluster of radical leaves, which are oblong and arrow-shaped at the base, with long stalks; stem-leaves few, on short stalks. Flowers in long terminal leafless clusters, at first greenish, but afterwards turning purplish-red. Abundant. Pastures between Meadfoot and Hope's Nose, Paignton, etc. (E. B. t. 137: not an accurate representation of the wild plant.) P. V.-VII.

10. **R. Acetosella** (*sheep's S.*)—In dry pastures. Plant varying from 3 or 4 inches to a foot high. Radical leaves linear-lanceolate and arrow-shaped at the base; stem-leaves narrow-lanceolate, but often some of them like the root-leaves. Flowers small, in slender terminal clusters, becoming, as is frequently the case with the whole plant, of a bright red colour. The leaves of this and those of the last described species are pleasantly acid. Common. Meadfoot cliffs. Warberry Hill, etc. (E. B. t. 1674.) P. V.-VII.

Ord. LXXIII. **THYMELACEÆ.**

DAPHNE. SPURGE-LAUREL.

D. Laureola (*common Spurge-Laurel.*)—In hedges, woods, and thickets. Stem from 1 to 3 feet high, stout, leafless below, but bearing at its summit a dense spreading cluster of large, smooth, lanceolate leaves, in the midst of which appear several axillary racemes of small green flowers. Leaves evergreen. Near Harford Bridge, on the Erme. Stoke Hill, near Exeter, *Fl. D.* (E. B. t. 119.) Sh. II.-IV.

Ord. LXXIV. SANTALACEÆ.

THESIUM. BASTARD TOADFLAX.

T. humile (*erect bastard Toadflax.*)—Stem erect, branched from the base; leaves fleshy, linear, 1-nerved, racemes spiked; flowers nearly sessile; fruit 4 or 5 times as long as the perianth. Mr. Babington states in his Manual that he found two specimens of this plant somewhere near Dawlish in 1829. P. VII. VIII.

Ord. LXXV. ARISTOLOCHIACEÆ.

Ord. LXXVI. EMPETRACEÆ.

Ord. LXXVII. EUPHORBIACEÆ.

MERCURIALIS. MERCURY.

M. perennis (*perennial, or dog's Mercury.*)—In woods and shady waste places. Stem a foot or more high; leaves stalked, oblong-lanceolate, serrated, situated mostly at the upper part of the stem; both male and female flowers in loose axillary spikes, green. Very common in woods and stony, bushy places. Park Hill wood. Carriage-drive to Torre Abbey. Petit Tor, etc. (E. B. t. 1872.) P. IV. V.

EUPHORBIA. SPURGE.

1. **E. Peplis** (*purple Spurge.*)—On seacoasts, in Devon and Cornwall. Stems numerous, procumbent and spreading, forked; leaves opposite, on short stalks, oblong-heartshaped; flowers single, axillary, small, with 4 pitted, yellow or red nectaries. Stems and stalks of a beautiful purplish-crimson; leaves of a greyish-green tinged with red. Sands between Torquay and Paignton. Goodrington sands. (E. B. t. 2002.) A. VII.-IX.

2. **E. helioscopia** (*sun S.*)—A very common weed in both waste and cultivated ground. Stem erect, round, branched slightly at the base; leaves obovate, membranous, slightly ser-

rate. Flowers in an umbel of 5 spreading rays, which are again divided into 3, and then into 2, yellowish-green. Whole plant full of an acrid milky juice. Abundant everywhere. (E. B. t. 883.) A. VI.-X.

3. **E. Paralias** (*sea S.*)—On sandy seacoasts. Not frequent. Root woody; stems about a foot high, with numerous barren leafy stems at the base; leaves leathery, elliptic-oblong, very closely packed together and lapping over each other. Umbel terminal, of 5 rays, each ray divided into 2. Floral leaves heart-kidney-shaped; flowers small, yellow. Coast about Torquay. Dawlish Warren. (E. B. t. 195.) P. VIII.-X.

4. **E. Portlandica** (*Portland S.*)—On the seacoast. Stem about 1 foot high, round and leafy, turning purplish in the Autumn. Leaves obovate-lanceolate, blunt, but with a small hair-like point at the summit. Umbel terminal, 5-rayed, with 2 or 3 smaller umbels beneath; flowers greenish-yellow. Meadfoot. Rocks by Livermead. Paignton. Teignmouth, etc. (E. B. t. 441.) P. V.-IX.

5. **E. Peplus** (*petty S.*)—In waste and cultivated ground. Stem upright, branching at the base, leafy. Leaves scattered, broadly obovate, stalked; umbel of 3 rays, repeatedly divided, bracts ovate; flowers small and green. Very common in fields and gardens. (E. B. t. 959.) A. VII.-IX.

6. **E. exigua** (*dwarf S.*)—In cornfields and waste places. The smallest of all the Spurges. Stem from 3 to 6 inches high, branched at the base; leaves linear, tapering to a point, alternate; umbel of 3 forked rays; bracts lanceolate; flowers small, greenish, with yellow nectaries. Common. (E. B. t. 1336.) A. VI.-IX.

7. **E. amygdaloides** (*wood S.*)—In woods, thickets, and under hedges. Stems red and tough, from 1 to 2 feet high, bare below but leafy above; leaves ovate-lanceolate, hairy beneath. Flower-stalks scattered, but the stem is crowned by a principal 5- or 6-rayed umbel; flowers small and yellow; floral leaves or bracts more or less tinted with rose-colour. Very abundant. (E. B. t. 256.) P. III.-V.

ORD. LXXVIII. **CALLITRICHACEÆ.**

CALLITRICHE. WATER STARWORT.

C. verna (*vernal Waterwort.*)—In pools, ditches, and slowly running water. Submerged leaves nearly all linear; floating leaves ovate, a bunch of them collected at the extremities of the branches, and spread out in rose-like form on the water; stems slender and hair-like; flowers white, in the axils of the upper leaves. Near Torre Abbey. Paignton osier-beds, etc. (E. B. t. 722.) A. IV.–IX. *Callitriche is the only genus in this Order.*

Ord. LXXIX. CERATOPHYLLACEÆ.

CERATOPHYLLUM. HORNWORT.

C. demersum (*common Hornwort.*)—In ditches, stagnant waters, and slow streams. Stem long and slender, floating; leaves in very close whorls, 2 or 3 times forked, segments linear; flowers sessile in the axils of the leaves; fruit armed with two lateral spines and crowned by the lengthened style. In the Clyst, by Clyst Bridge, *Fl. D.* (E. B. t. 947.) P. VI. VII. *This Order contains only one genus.*

Ord. LXXX. URTICACEÆ.

Subord. I. *URTICEÆ.*

URTICA. NETTLE.

1. **U. urens** (*small Nettle.*)—In waste places and under hedges, too frequent. Stem 12 or 18 inches high; leaves opposite, elliptical, deeply serrate, bright green, covered with minute hair-like bristles; flowers in axillary spikes. Abundant. (E. B. t. 1236.) A. VI.–IX.

2. **U. dioica** (*great N.*)—Same situations as the last. Stem erect, from 2 to 3 feet high, bluntly 4-cornered, bristly; leaves opposite, stalked, heart-shaped at the base, pointed at the extremities, sharply serrate; but sometimes ovate-lanceolate and rounded at the base, covered with stinging hairs; flower-spikes in pairs, axillary. Everywhere abundant. (E. B. t. 1750.) P. VII.–IX.

PARIETARIA. PELLITORY OF THE WALL.

P. officinalis (*common Pellitory.*)—On old walls and among stony rubbish. Plant from 6 to 10 or 12 inches high, having much the appearance of a Nettle; stem branched and spreading, tinged with red, leaves ovate-lanceolate, entire, slightly hairy, stalked. Flowers in sessile clusters at the base of the leaves, reddish. For an account of the structure of the flowers, which is very curious, see Hooker and Arnott's 'British Flora.' Old walls about Torre Abbey. Cockington, etc. (E. B. t. 879.) P. vi.-ix.

SUBORD. II. *CANNABINEÆ.*

HUMULUS. HOP.

H. Lupulus (*common Hop*).—In hedges and thickets. Male and female flowers on separate plants. Stems twining and twisting over bushes and small trees, often to a great height. Leaves opposite, stalked, large, from 3- to 5-lobed, heart-shaped at the base, sharply serrate. On the male plant, flowers in loose clusters in the axils of the upper leaves, yellowish-green; on the female, in axillary roundish heads or spikes of thickly crowded bracts or scales, with 2 sessile flowers in each axil: these heads or catkins are the parts of the plant so extensively used in the manufacture of beer. Male plant, in a hedge on the road between Churston Ferrers and Brixham. Female plant, Marychurch. (E. B. t. 427.) P. vii. viii.

ORD. LXXXI. ULMACEÆ.

ULMUS. ELM.

1. **U. montana** (*Wych Elm*).—In woods and hedges. A large and exceedingly picturesque tree, with wide and spreading branches. Leaves nearly sessile, broadly ovate, doubly serrate, unequal at the base, downy beneath. Flowers reddish-purple, in thick clusters, fruit broadly ovate, or roundish, green, somewhat like a hop-scale. Common in woods and hedges. (E. B. t. 1887, and *U. major*, t. 2542.) T. iii. iv.

2. **U. campestris** (*common Elm*).—In hedges, etc. Not unlike the last, but generally a taller and straighter-growing tree; leaves not quite so large; but it is mainly distinguishable by its fruit, which is oblong and deeply notched; its characteristics however are not constant, and the different varieties of Elm seem to glide into each other. Common. (E. B. t. 1886. *U. suberosa*, t. 2161, and *U. glabra*, t. 2248.) T. iii.-v.

ORD. LXXXII. ELÆAGNACEÆ.

ORD. LXXXIII. MYRICACEÆ.

MYRICA. GALE.

M. Gale (*sweet Gale, or Dutch Myrtle.*)—In bogs and moory situations. Diœcious. Plant from 2 to 3 feet high; leaves lanceolate, serrated towards the top. Catkins sessile towards the ends of the branches; stamens of male flowers from 4 to 8; female catkins shorter and rounder than those of the male, with the long styles projecting from between the scales. The plant is very fragrant when bruised. Milber Down. Forde bog. Bovey Heath. Near Chagford, etc. (E. B. t. 562.) Sh. v.–vii.

ORD. LXXXIV. BETULACEÆ.

BETULA. BIRCH.

B. alba (*common Birch.*)—In woods and hedges. Monœcious. A slender and graceful tree, with beautifully drooping branches, bark of the stem white, easily peeling off; leaves ovate and acute, irregularly serrate. Catkins pendulous; flowers of the male with from 8 to 12 stamens; female catkin compact, with 3 flowers to each scale; seed winged. Park Hill wood, etc. (E. B. t. 2198.) T. iv. v.

ALNUS. ALDER.

A. glutinosa (*common Alder.*)—Wet meadows and by sides of streams. A middling-sized tree, with crooked, spreading branches. Leaves alternate, stalked, broadly oval or roundish, waved and serrated, downy beneath at the divisions of the veins. Male catkins long and drooping; flowers with 4 stamens; fertile ones small and oval, with reddish scales. Very common. (*Betula Alnus*, E. B. t. 1508.) T. iii. iv.

Ord. LXXXV. SALICACEÆ.

SALIX. WILLOW, SALLOW, OSIER.

Mr. Bentham, in his introduction to the Willow tribe, says, "The great variations in the shape of the leaves of many species, and the difficulty of matching the male and female specimens, or the young and old leaves of those species which flower before the leaves are out, have produced a multiplication of supposed species, and a confusion in their distinction, beyond all precedent." In his Handbook he reduces the species of British Willows to 15, considering that number to include all that can be called truly distinct species.

1. **S. purpurea** (*purple Willow.*)—In marshy places and on banks of rivers. Either a decumbent shrub or a small tree. Leaves lanceolate, widening upwards, pointed, narrowing below into the stalk, serrate, green on the upper but whitish or silky on the under side. Catkins appearing before the leaves, the males sessile, anthers purple; females shortly stalked, with 2 or 3 small leaf-like bracts at their base; capsule sessile and very cottony. Bradley, near Newton. Banks of the Teign, near Whyddon Park. (E. B. t. 1388.) Bentham gives, as varieties of this, *S. Helix*, E. B. t. 1343; *S. Forbyana*, t. 1344; *S. rubra*, t. 1145; and *S. Lambertiana*, t. 1359. T. III. IV.

2. **S. triandra** (*blunt-stipuled triandrous W.*)—In wet and marshy places, willow-beds, etc. Height from 20 to 30 feet; leaves oblong-lanceolate, pointed, white underneath but not downy. Catkins cylindrical, on short leafy shoots; males with 3 stamens beneath each scale; scales of female catkins persistent. Paignton. Totness, etc. (E. B. t. 1435, and *S. amygdalina*, t. 1936.) T. IV. V.

3. **S. fragilis** (*crack W.*)—In wet woods and willow-beds. A large and bushy tree, with crooked branches. Leaves broadly lanceolate, pointed, large, of a dark shining green, finely serrate. Catkins long and loose; male flower with 2 stamens; capsule of the female tapering at the top. Copse at Chudleigh. Marshes below Clyst Bridge. Banks of the Teign, near Chagford. (E. B. t. 1807.) The varieties of this, frequently met with in the same localities, are *S. Russelliana*, E. B. t. 1808, and *S. decipiens*, t. 1937. T. IV. V.

4. **S. alba** (*common white W.*)—In moist woods, meadows, hedgerows; also in marshes and by riversides. A large tree, growing from 50 to 80 feet high; foliage of an ashy-grey. Leaves narrow-lanceolate, serrate, the lowest serratures glandular, silky on both sides. Catkins loose, on short, lateral, leafy stalks. Sta-

mens 2; capsule nearly sessile, ovate, but slightly tapering at the top. Frequent in damp woods. (E. B. t. 2430.) Also the variety with leaves less silky beneath (E. B. t. 2431). T. v.

5. **S. vitellina** (*yellow W., or golden Osier.*)—In hedges and osier-grounds. Placed by Babington as a variety of *S. alba*, from which it differs in its bright yellow branches, in its long capsule-scales, and its shorter and broader leaves; the figure in 'English Botany' represents the leaves narrower. Paignton osier-beds. (E. B. t. 1389.) T. v.

6. **S. fusca** (*dwarf silky W.*)—In moist and dry heaths. A low and straggling shrub, 12 or 18 inches high; stem decumbent at the base, then ascending and much branched; the foliage and new shoots very white and silky. Leaves either oblong-lanceolate, or narrow-oblong, or ovate, slightly serrate. Catkins roundish, sessile when in flower, accompanied by a few leafy bracts; fruiting catkins longer and slightly stalked; capsules silky. Forde bog. Bovey Heath. Heathy places at Lindridge. (E. B. t. 1960; and the varieties *S. repens*, E. B. t. 183; *S. argentea*, t. 1364; *S. arbuscula*, t. 1366; *S. prostrata*, 1959; *S. parviflora*, t. 1961; and *S. adscendens*, t. 1962.) Sh. III. IV.

7. **S. ambigua** (*ambiguous Willow.*)—Similar situations to the last. Given by Bentham as a variety of the last. Its habit rather less creeping, and distinguished by its oval, obovate, or hairy lanceolate leaves, which are somewhat wrinkled. Forde bog. Bovey Heath. (E. B. S. t. 2733.) Sh. v.

8. **S. viminalis** (*common Osier.*)—In wet meadows and osier-beds. From 10 to 20 feet high, with long switchy branches. Leaves linear or linear-lanceolate, sometimes 4 or 5 inches long, silky and shining on their under sides. Catkins nearly if not quite sessile; males with 2 stamens; capsules of female downy, tapering upwards. Paignton osier-beds, and watery meadows. (E. B. t. 1898.) *S. stipularis*, t. 1214, and *S. Smithiana* (*S. mollissima*, E. B. t. 1509), appear to be varieties of this. Sh. IV. V.

9. **S. cinerea** (*grey Sallow.*)—In wet woods and hedges, by the sides of rivers, and in swampy places. Sometimes merely a bushy shrub, but in sheltered situations rising to a tree of from 20 to 30 feet high. Leaves elliptical or obovate-lanceolate, serrated, downy beneath and reticulated with rather prominent veins. Catkins sessile; capsules lanceolate, pointed, and silky. Very common. (E. B. t. 1897; *S. aquatica*, t. 1437; and *S. oleifolia*, t. 1402.) T. or Sh. III. IV.

10. **S. aurita** (*round-eared Sallow.*)—In damp woods, etc. A small bushy shrub, from 3 to 4 feet high. Leaves about an inch long, downy beneath, varying from obovate to roundish or

oblong, obtuse, but terminating in a curved point, easily distinguished by their greatly wrinkled appearance. Male catkins small and closely sessile; female catkins on short stalks, with small leafy bracts; capsules stalked, downy. Warberry Hill. Hedges by the Paignton road. Copse by the brook at Chudleigh. (E. B. t. 1487.) Sh. IV. V.

11. **S. caprea** (*great round-leaved Sallow.*)—In dryish woods and hedges. A tall shrub or moderate-sized bushy tree, varying from 15 to 30 feet in height. Leaves often large, ovate or oblong, pointed, distinctly serrate, downy beneath, rounded or broadly heart-shaped at the base. Catkins sessile; the males large and handsome, of an oval form, with bright yellow anthers; the female catkins broad and short, with leafy bracts; capsule swelling below, and tapering towards the top, downy-white. Common in woods and hedges. (E. B. t. 1488.) The variety *S. sphacelata*, in a hedge by the Meadfoot road. (E. B. t. 2333.) Bentham thinks, that probably *S. acuminata* (E. B. t. 1434) belongs also to this. T. IV. V.

12. **S. nigricans** (*dark-leaved Sallow.*)—By sides of streams and in osier-grounds. Leaves elliptic-lanceolate, crenate, but very variable in size and shape; young shoots very hairy towards the summit; the leaves always turn black by being pressed and dried. Catkins about an inch in length, yellow; capsules stalked, conical, and pointed at the top. Near Berry Pomeroy Castle. (E. B. t. 1213.) Sh. IV.-VI.

POPULUS. POPLAR.

1. **P. alba** (*great white Poplar, or Abele.*)—By the sides of streams and in moist open woods. A large, handsome tree, with smooth greyish bark and wide spreading branches; the young shoots and under sides of the leaves covered with a thick white down. Leaves roundish heart-shaped, irregularly toothed, those of the young shoots 5-lobed, somewhat palmate. Male catkins long, cylindrical, and drooping, anthers violet-blue; female catkins shorter, ovate. Frequent in woods and plantations. (E. B. t. 1618.) *P. canescens* (the grey Poplar) is a variety with smaller leaves, which are undivided, and not so white beneath. T. III. IV.

2. **P. tremula** (*trembling P., or Aspen.*)—In woods, etc. A smaller and much more slender tree than the last. Leaves nearly round, toothed, pale beneath, but not downy; leaf-stalks laterally compressed, allowing the leaf to wave and quiver with the slightest breath of wind. Catkins pendulous, much smaller than

those of *P. alba.* Woods, etc. Chudleigh. (E. B. t. 1909.) T. III. IV.

3. **P. nigra** (*black P.*)—In moist places and by sides of rivers; not truly indigenous. A tall tree, growing in a pyramidal form. Leaves nearly triangular, tapering to a point, serrate; catkins long, loose, and drooping; stamens more numerous than in the other species, varying from 8 to 30. Very frequent. (E. B. t. 1910.) T. III.

Ord. LXXXVI. CUPULIFERÆ.

FAGUS. BEECH.

F. sylvatica (*common Beech.*)—In woods and plantations. Male and female flowers on the same tree. A large and beautiful tree, with smooth round trunk and thickly-clustered branches. Leaves on short stalks, ovate, obscurely toothed with fine hairs on the edges. Male flowers in a roundish catkin of about a dozen flowers; female catkins globular, with 2 or 3 sessile flowers in their centres. Nuts 2 or 3, contained in a hard spiny involucre. Park Hill wood. Ansti's Cove lane. Side of the Newton road, near Milber Down. Holy Street, near Chagford, very fine examples. (E. B. t. 1846.) T. III. IV.

CASTANEA. CHESTNUT.

C. vulgaris (*Spanish Chestnut.*)—In woods, etc. A stately and magnificent tree, but by most botanists not considered a native. Height from 50 to 80 feet. Branches wide and spreading; leaves large and shining, oblong-lanceolate, sharply serrate. Barren flowers in long, pendulous, interrupted spikes; fertile flowers from 1 to 3, in a 4-lobed involucre; nuts 1 or 2 in the greatly enlarged calyx, which is invested with tangled and complicated prickles. In woods and plantations about Torquay. Churston Ferrers, by the roadside; a very large tree. (*Fagus Castanea*, E. B. t. 886.) T. V.

QUERCUS. OAK.

Q. Robur (*common British Oak.*)—In woods and hedges. A bold and stately tree, attaining to the greatest age of any of our native trees. Leaves obovate-oblong, deeply sinuate, lobes blunt. Male flowers in slender pendulous clusters; fruit clustered or spiked, on a fruit-stalk from $\frac{1}{4}$ to $\frac{1}{2}$ inch long. The variety *Q. sessiliflora* (E. B. t. 1845) has the fruit-clusters sessile. Very abundant in woods, etc. (E. B. t. 1342.) T. IV. V.

CORYLUS. HAZEL-NUT.

C. Avellana (*common Hazel.*)—In hedges and thickets. A large shrub, but sometimes almost a tree. Leaves roundish heart-shaped, lobed, and pointed, serrated, downy on both sides. Male catkins long and pendulous; females sessile on the sides of the twigs, with protruding crimson styles; nut ovate, 1-celled, covered by the leathery-looking fimbriated calyx. Very common. (E. B. t. 723.) Sh. III. IV.

CARPINUS. HORNBEAM.

C. Betulus (*common Hornbeam.*)—In woods and hedges. A small tree, rarely reaching the height of 30 feet. Leaves alternate, ovate or heart-shaped, acute, doubly serrate, stalked; when young curiously plaited. Male catkins sessile; female flowers in loose, terminal, bracteated clusters; scales of the fruit in 3 segments, the middle one the longest; nut small, ovoid, striated. Chagford. (E. B. t. 2032.) T. V.

ORD. LXXXVII. **CONIFERÆ**.

Pinus sylvestris (Scotch Fir), belonging to this Order, has been extensively planted in this neighbourhood; and *Taxus baccata* (common Yew) is also to be found in many localities within my prescribed circuit. An ancient Yew-tree formerly grew in the Waldon Hill wood, near the end of the Rock Walk, but was cut down to make way for the recent buildings there; these trees, however, not being indigenous to the county, cannot, of course, be legitimately placed in our Flora.

CLASS II. **MONOCOTYLEDONOUS** OR **ENDOGENOUS FLOWERING PLANTS.**

SUB-CLASS I. **PETALOIDEÆ.** (ORD. LXXXVIII.–CV.)

ORD. LXXXVIII. **HYDROCHARIDACEÆ.**

HYDROCHARIS. FROG-BIT.

H. Morsus-ranæ (*common Frog-bit.*)—In ponds and ditches. Stems floating, having the appearance of creeping runners, and sending down long radicals. Leaves stalked, roundish kidney-shaped, entire. Flowers white and delicate, barren and fertile ones on different plants; male flowers with from 9 to 12 stamens, females with 6 styles, deeply 2-cleft; capsules 6-celled, containing numerous seeds, which are covered with prominent spirally twisted cells. Ditches at Exminster and Powderham. (E. B. t. 808.) P. VII. VIII.

ORD. LXXXIX. **ORCHIDACEÆ.**

EPIPACTIS. HELLEBORINE.

E. latifolia (*broad-leaved Helleborine.*)—In woods and shady lanes, etc. Root creeping; stem from 1 or 2 to 3 feet high; lower leaves ovate, very broad, upper ones narrower, lanceolate; all the leaves strongly ribbed. Flowers green, with a purple lip, sometimes all purple, pendulous, in a long lateral raceme; petals shorter than the ovate-lanceolate sepals, lip small. Cockington. Chudleigh. Near Haldon House, *Fl. D.* (*Serapias*, E. B. t. 269.) P. VII. VIII.

LISTERA. BIRD'S-NEST, TWAYBLADE.

1. **L. ovata** (*common Twayblade.*)—In damp woods, moist meadows, and wet places. Stem about 1 foot high, with 2 op-

posite ovate leaves. Flowers in a long lax spike, crowning the naked stem, small and greenish; lip long and 2-cleft. Wood by the Newton road. Paignton osier-beds. Berry Pomeroy wood. *Ophrys*, Linn. (E. B. t. 1548.) P. v. VI.

2. **L. Nidus-avis** (*common Bird's-nest.*)—In moist woods and thickets. Root of numerous, crowded, fleshy, and bulbous fibres; stem a foot or more high, clothed with sheathing scales, and bearing no leaves. Flowers of a dusky brown, in a terminal spike; lip oblong, with 2 spreading lobes. Whole plant of a dingy reddish-brown, having, at first sight, much the appearance of an *Orobanche*. Berry Pomeroy woods. Ugbrooke Park. *Ophrys*, Linn. *Neottia*, Bab. (E. B. t. 48.) P.? VI.

NEOTTIA. LADY'S-TRESSES.

N. spiralis (*fragrant Lady's-Tresses.*)—On dry hilly pastures. Root with 2 to 4 oblong root-knobs. Leaves radical and spreading, oblong or broadly ovate; stem from 4 to 6 inches high, with short sheathing bracts. Flowers in a spiral terminal spike, greenish-white, almond-scented. Park Hill, near the stone seat. Daddy-hole Plain. Babbicombe Down. *Spiranthes autumnalis*, Bab. and Benth. (E. B. t. 541.) P. VIII.-IX.

ORCHIS. ORCHIS.

1. **O. Morio** (*green-winged meadow Orchis.*)—In pastures and meadows. Root-knobs 2, undivided. Stem from 6 to 8 or 9 inches high; leaves almost radical, lanceolate. Flowers few, in a loose spike, purple; bracts slender and tinged with pink; outer sepals purplish, arching over the petals; lip 3-lobed, pale in the middle, with purplish spots; spur blunt, not quite so long as the seed-vessel. Daddyhole Plain. Ilsham Down. Warberry Hill. (E. B. t. 2059.) P. v. VI.

2. **O. mascula** (*early purple O.*)—In woods and pastures. Root-knobs 2, undivided. Stem a foot or more high. Leaves chiefly radical, long-lanceolate, generally more or less spotted with purple; flowers in a loose spike, numerous and showy, generally purple, but sometimes flesh-coloured or white, bracts coloured, the 2 outer sepals turned upwards and converging; lip of the corolla 3-lobed, the middle lobe largest and notched; spur ascending, longer than the germen. Daddyhole Plain. Cliff walks at Meadfoot, and beyond Hope's Nose. Bradley woods, etc. (E. B. t. 631.) P. v.

3. **O. ustulata** (*dwarf dark-winged O.*)—On dry, hilly, lime-

stone pastures. A beautiful little Orchis, from 4 to 6 inches high. Knobs of the roots entire. Leaves lanceolate, acute. Flowers in a dense, oblong, terminal spike, small and numerous; lip of the corolla white, with raised purple spots, 3-partite, with linear-oblong lobes. Helmet dark reddish-purple. Spur one-third the length of the germen. Daddyhole Plain (nearly extinct). Babbicombe Down. Berry Head. (E. B. t. 18 : too dingy.) P. VI.

4. **O. maculata** (*spotted palmate O.*)—Wet pastures, woods, and heaths. Root-knobs 2, palmate. Stem about a foot high. Lower leaves blunt, upper linear-lanceolate, all usually spotted with purple. Flowers white or light purple, with darker purple streaks; lip flat, deeply 3-lobed, middle lobe longer and narrower than the lateral ones; spur shorter than the germen, bracts as long as or exceeding it. Orchard by the Babbicombe road. Wood on the Newton road. Cockington. Osier-beds at Paignton. Wet wood at Lindridge. Forde bog. (E. B. t. 632.) P. V. VI.

5. **O. latifolia** (*marsh O.*)—In marshes and damp meadows. Root-knobs 2, palmate. Stem from 12 to 18 inches high, hollow. Leaves large and lanceolate, spreading. Flowers in a dense spike; bracts long, lower ones exceeding the flowers; spur shorter than the germen; lip indistinctly 3-lobed, its sides reflexed and wavy; flowers varying in colour from pale pink to purple; the lip spotted and streaked with purple lines. Paignton osier-beds. Meadows at Shiphay. Forde bog. Banks of the Dart, near Totness, etc. (E. B. t. 2308.) P. VI.

6. **O. pyramidalis** (*pyramidal O.*)—In pastures on a limestone soil. Root-knobs 2, undivided. Stem a foot or more high. Leaves linear-lanceolate, very acute. Flower-spike pyramidal or cylindrical; flowers bright reddish-purple, sometimes white; lip equally 3-lobed; spur thread-like, longer than the germen. Daddyhole Plain. Babbicombe Down. Goodrington. Kerswell Down. Chudleigh Woods. (E. B. t. 110.) P. VII.

HABENARIA. HABENARIA, BUTTERFLY ORCHIS.

1. **H. bifolia** (*lesser Butterfly-Orchis.*)—In moist thickets, meadows, and marshes. Root-knobs tapering, undivided. Stem about 1 foot high. Root leaves 2, large, oblong or elliptical; narrow and bract-like. Flowers white, in a loose spike; spur slender, twice as long as the germen; lip linear, entire. Fir wood on the Braddons Hill (extinct). Cockington lanes. Wood on the Newton road. Chudleigh. (E. B. S. 2806.) P. VI. VII.

2. **H. chlorantha** (*great B.*)—Similar situations to the last. Much taller and stouter than the last, which it greatly resembles, and of which it is most probably but a luxuriant variety, the chief difference, except that of size, being that in this the anther is more dilated at the base than it is in *H. bifolia*. Wood on the Newton road. (*Orchis bifolia*, E. B. t. 22.) P. VI. VII.

OPHRYS. OPHRYS.

O. apifera (*Bee-Ophrys.*)—On limestone pastures and chalky soils. Root-knobs roundish, undivided. Stem from 12 to 18 inches high; leaves broad. Flowers few, large, and distant, resembling somewhat a bee upon the wing; lip swelling, 5-lobed, the 2 lower ones prominent and with a hairy base, the intermediate ones reflexed. Sepals whitish or greenish, tinged with pink; lip velvety, rich brown variegated with yellow; anther lengthened out, terminating in a hooked point. Daddyhole Plain, on the slope leading to the Quarry. Babbicombe Down. Berry Head. (E. B. t. 383.) P. VI. VII.

ORD. XC. **IRIDACEÆ.**

IRIS. IRIS, FLOWER-DE-LUCE.

1. **I. Pseudacorus** (*yellow water Iris, or Cornflag.*)—In watery places, wet meadows, and woods. Root large; stem roundish; leaves long and sword-shaped. Flowers large, bright yellow; stigmas 3-parted, having the appearance of petals, covering the stamens; seed-vessel oblong, 3-cornered. Very common. Torre Abbey meadow. Paignton. Forde bog. Kingskerswell, etc. (E. B. t. 578.) P. VI. VII.

2. **I. fœtidissima** (*stinking Iris.*)—In woods and on hedgebanks. Stem flattened; leaves sword-shaped, narrower than those of *I. Pseudacorus*, yielding a fetid odour when bruised. Flowers of a dull pale blue, petals narrow; seed-vessel large, containing, when ripe, numerous bright red seeds. Very abundant in all our lanes, and in every wood and thicket. (E. B. t. 596.) P. VI. VII.

TRICHONEMA. TRICHONEMA.

T. Columnæ (*Columna's Trichonema.*)—In sandy places. Root bulbous; plant 3 or 4 inches high; leaves slender and thread-like, furrowed, longer than the flower-stalks. Flowers solitary, pale violet-colour, yellow a*·* the lower part within. Dawlish Warren. (*Ixia Bulbocodium*, E. B. t. 2549.) P. III. IV.

Ord. XCI. AMARYLLIDACEÆ.

NARCISSUS. NARCISSUS, DAFFODIL.

1. **N. Pseudo-Narcissus** (*common Daffodil.*)—In damp woods and thickets. Root bulbous. Leaves linear, blunt at the end. Flower-stalk single-flowered; flowers large, yellow; petals 6; nectary bell-shaped, crisped at the margin, as long as the petals; stamens 6, within the nectary. West Hill, near Torquay. Shiphay. Kingskerswell. Barton. (E. B. t. 17.) P. III. IV.

2. **N. biflorus** (*pale Narcissus.*)—In sandy fields. Root bulbous. Leaves linear, obtuse, keeled. Flower-stalk usually bearing 2 flowers within the spatha; flowers of 6 pale sulphur-coloured spreading petals; nectary short, bright yellow, with a white crenated margin. Shiphay. Barton. Fields at Paignton, near the Dartmouth road. Fields by the Dart, above Totness Bridge. (E. B. t. 276.) P. IV. V.

GALANTHUS. SNOWDROP.

G. nivalis (*common Snowdrop.*)—In old orchards, meadows, and thickets. Root bulbous; leaves 2, broadly linear. Flowers solitary, white, drooping, inner segments greenish. Orchards at Marychurch. Field at Barton Ridge. Near the brook at Chudleigh. (E. B. t. 19.) P. II. III.

Ord. XCII. DIOSCOREACEÆ.

TAMUS. BLACK-BRYONY.

T. communis (*common Black-Bryony.*)—In hedges and thickets. Root large, thick and fleshy, acrid, somewhat like a Yam in appearance, and abounding in starch. Stems very long and climbing; leaves heart-shaped, acute, undivided. Flowers greenish-white, on long-stalked axillary racemes. Berries red. Hedges about Torquay. Walks above Hope's Nose. Wood near Ansti's Cove. Hedges at Shiphay. Wood by the Newton road, etc. (E. B. t.91.) P. V. VI.

Ord. XCIII. TRILLIACEÆ.

Ord. XCIV. LILIACEÆ.

* *Roots never bulbous.*

ASPARAGUS. ASPARAGUS.

A. officinalis *(common Asparagus.)*—Seacoast and banks of rivers. Rare. Root creeping and matted; stem erect, much branched, from 1 to 2 feet high; leaves numerous, long and hair-like, in feathery clusters. Flowers small, greenish-white, 2 or 3 together, on slender axillary flower-stalks; berries globular, bright red. Banks of the Exe, near Topsham, *Fl. D.* Near the bridge above Exeter, *Mr. Earle.* (E. B. t. 339.) P. VIII.

RUSCUS. BUTCHER'S-BROOM.

R. aculeatus *(common Butcher's-Broom.)*—In woods and thickets. Plant evergreen, from 2 to 3 feet high; stem erect and branching; leaves numerous, rigid, ovate, terminating in a sharp spinous point. Flowers solitary, small and white, on the upper surface of the leaves, which are curiously turned round by a twist at their base. Berries bright red, large in comparison with the flower. Cockington wood. Bradley, near Newton. Wood at Goodrington. (E. B. t. 560.) Sh. III.-V.

** *Root bulbous.*

AGRAPHIS. BLUEBELL.

A. nutans *(wild Hyacinth, or Bluebell.)*—In woods, thickets, and on hedge-banks. Leaves linear, channelled; flower-stem a foot or more high, terminating in a beautiful nodding cluster of bell-shaped blue flowers, each flower accompanied by a small, narrow bract at the base of the flower-stalk. Very abundant. *Scilla*, Sm. (E. B. t. 377.) *Hyacinthus nonscriptus*, Linn. *Endymion*, Bab. P. v.

ALLIUM. ONION, GARLIC.

1. **A. vineale** (*crow Garlic.*)—In dry fields and waste ground. Root-bulb ovate. Leaves few, long, hollow, and curling. Umbel bearing small greenish bulbs mixed with the flowers, or bulbs only; flowers few, pale red. Park Hill. Daddyhole Plain. Cliffs near Hope's Nose. Ilsham Down. (E. B. t. 1974.) P. VII.

2. **A. ursinum** (*broad-leaved G. or Ramsons.*)—In moist woods and shady hedge-banks. Leaves nearly all radical, stalked, ovate-lanceolate, long, flat, and spreading, bright green; flower-stem from 8 to 10 inches high, crowned by a loose umbel of numerous, pretty, white flowers. Whole plant has a strong odour of garlic when bruised. Torre Abbey drive. Cockington lanes: Upton lanes. Bradley woods, in great abundance. (E. B. t. 122.) P. V. VI.

SCILLA. SQUILL.

S. autumnalis (*autumnal Squill.*)—In dry pastures. Bulb large, coated. Flower-stems from 4 to 8 inches high, appearing after the linear leaves have withered away, according to Bentham; but coming before the leaves show themselves, according to Babington; flowers small, bright pinkish-purple. Park Hill, near the stone seat. Daddyhole Plain. Warberry Hill. (E. B. t. 78.) P. VII. VIII.

ORD. XCV. MELANTHACEÆ.

COLCHICUM. MEADOW-SAFFRON.

C. autumnale (*common Meadow-Saffron.*)—In wet meadows and pastures. Root large and tuberous. Flowers several, bright purple, appearing before the leaves. Leaves flat and lanceolate, often a foot long, and as much as an inch or an inch and a half broad, appearing in the spring together with the capsule which then appears above the surface of the ground, the leaves soon afterwards withering away. Formerly in Torwood meadows, the site of the present public gardens. (E. B. t. 133.) P. IX. X.

PETALOIDEÆ. 131

ORD. XCVI. **RESTIACEÆ.**

ORD. XCVII. **JUNCACEÆ.**

JUNCUS. RUSH.

a. *Without leaves. The barren stems resembling leaves.*

1. **J. effusus** (*soft Rush.*)—In marshy ground, common. The creeping rootstock throws up thick tufts of leafless stems from 2 to 4 feet high, which are faintly striated and soft, some of which bear a few inches from the top a lateral, loose, and spreading panicle of greenish or brown flowers. *J. conglomeratus* (common Rush), E. B. t. 835, very much resembles this, but bears its flowers in a dense, globose, and compact panicle. Bentham unites them under the name of *J. communis*. Forde bog. Meadow by the Babbicombe road, opposite the Lower Warberry road. Bovey Heath, etc. (E. B. t. 836.) P. VII.

2. **J. glaucus** (*hard R.*)—In wet pastures and by roadsides. Very near the last in appearance, but much shorter, harder, and stiffer, having the pith interrupted while in the two others the pith is continuous; panicle loose and much branched; flowers pale brown; fruit black. Fields about Cockington. Barton. Forde bog, etc. Common. (E. B. t. 665.) P. VII.

3. **J. maritimus** (*lesser sharp Sea-R.*)—In sandy marshes. Stems in tufts, from 1 to 3 feet high, sharp-pointed, as are also the sheathing scales at their base; panicle long and loose, consisting of numerous flowers in little distinct clusters. Paignton Marsh. Starcross. Marshes at Kingsteignton. (E. B. t. 1725.) P. VII. VIII.

4. **J. acutus** (*great sharp Sea-R.*)—On sandy seashores. Very much taller and stouter than the last, as well as more rigid and prickly; stems from 3 to 6 feet high. Panicle dense and compact; flowers brown; fruit-capsules very large, protruded, of a rich glossy brown. Goodrington marshes. (E. B. t. 1614.) P. VII. VIII.

b. *Stems leafy.*

5. **J. acutiflorus** (*sharp-flowered jointed R.*)—In boggy places. Stem from 18 inches to 2 feet high; leaves 3 or 4, slightly compressed, and having a jointed appearance. Flowers from 3 or 4 to 6 in a cluster, greenish-brown, arranged in a compound terminal panicle. Capsule pale brown. Paignton. Goodrington. Blagdon. (*J. articulatus*, E. B. t. 238.) P. VI.-VIII.

Bentham places *J. lamprocarpus*, E. B. t. 2143, with this, and thinks also that *J. obtusifolius* should probably be added. Forde bog.

6. **J. uliginosus** (*lesser bog jointed R.*)—In boggy and swampy places. Very variable in appearance; stems 3 or 4 inches high, either erect or decumbent; leaves slender, tapering and pointed, radical ones with broad red sheaths. Panicle irregular, nearly simple, with few distant clusters of flowers. Forde bog. Bovey Heath. *J. supinus*, Bab. E. B. t. 801. The var. β of Babington, *J. nigritellus*, with the capsule shorter, and the filaments of the stamens nearly twice as long as the anthers; in a bog at Ivybridge. P. VI.-VIII.

7. **J. castaneus** (*clustered alpine R.*)—In mountainous bogs chiefly. Rare. Root slightly creeping. Stem from 8 to 12 inches high, hollow, bearing 2 or 3 channelled leaves. Flower-heads terminal, single or 2 or 3 together, 4- or 5-flowered. Capsules ovate-oblong, chocolate-coloured. Bovey Heath, *Mr. C. Parker*. P. VII. VIII.

8. **J. compressus** (*round-fruited R.*)—In damp marshy places. Stems upright and rather slender, from 12 to 18 inches high, more or less compressed at the base; leaves nearly radical, few and shorter than the stem, channelled and narrow. Flowers in a loose terminal panicle, bright brown. Capsules roundish-ovate. Bovey Heath. (E. B. t. 934.) P. VI.-VIII.

9. **J. bufonius** (*toad R.*)—In wet and marshy places. Stems numerous, forming thick tufts from 2 to 8 inches high, branching in a forked manner; leaves slender and thread-like. Flowers solitary, scattered, mostly sessile, green, with white edges to the perianth-segments. Chudleigh. Bradley. E. B. t. 802. A. VII. VIII.

10. **J. squarrosus** (*heath R.*)—On wet heaths and moors. Whole plant stiff and rigid, from 6 to 12 inches high; leaves numerous, radical, in spreading tufts, linear and channelled. Flowers in a terminal, compound, slightly branched panicle, large, dark brown. Capsule pale brown, shining. Milber Down. Bovey Heath. (E. B. t. 933.) P. VI. VII.

LUZULA. WOOD-RUSH.

1. **L. sylvatica** (*great hairy Wood-Rush*.)—In woods and thickets, in hilly districts. The largest of this genus, the stems rising from 1 to nearly 3 feet high; leaves large, flat, striated, and grass-like, fringed with long white hairs. Flowers in clusters of 2 or 3, in a loose, spreading, compound panicle. Berry

Pomeroy woods. Banks of the Teign at Gidleigh, near Chagford. Holne Chase. Banks of the Exe. (*Juncus*, E. B. t. 737.) P. IV.-VI.

2. **L. pilosa** (*broad-leaved hairy W.*)—In woods, thickets, and hedges. Stems slender, from 6 to 12 inches high; leaves lanceolate, hairy, chiefly radical. Flowers in an irregular terminal panicle, the flower-stalks of which are from 1- to 3-flowered, when in threes the central flower nearly sessile, the two others on slender stalks. Common in the situations above stated. (*Juncus*, E. B. t. 736.) P. V.

3. **L. Forsteri** (*narrow-leaved hairy W.*)—In woods and thickets. More slender than the last; stems about 1 foot high; leaves linear, hairy. Panicle terminal, slightly branched, flower-stalks 1-flowered, erect both in flower and fruit. Berry Pomeroy woods. (*Juncus*, E. B. t. 1293.) P. V.

4. **L. campestris** (*field W.*)—In heathy places, woods, and dry pastures. Stems from 4 inches to a foot high; leaves linear, hairy. Panicle in 3 or 4 compact ovate heads, with 6 or 8 flowers in each. Walks above Meadfoot. Cliff walks between Hope's Nose and Ansti's Cove. Wood near Bishop's Stowe. Bradley Wood, etc. (*Juncus*, E. B. t. 672.) P. IV. V.

NARTHECIUM. BOG-ASPHODEL.

N. ossifragum (*Lancashire Bog-Asphodel*.)—In boggy places and moors. Stem erect and rigid, from 6 to 8 inches high, slightly leafy. Leaves mostly radical, shorter than the stem, linear sword-shaped, in 2 opposite ranks. Flowers in a terminal, elongated raceme, bright yellow. Forde bog. Bovey Heath. Boggy places in Dartmoor. Haldon. (E. B. t. 535.) P. VI. VII.

ORD. XCVIII. BUTOMACEÆ.

BUTOMUS. FLOWERING-RUSH.

B. umbellatus (*common Flowering-Rush.*)—In rivers, ditches, and ponds. Leaves all radical, linear, erect, and triangular; flower-stalk springing from the root, much taller than the leaves, stout and rush-like, bearing an irregular umbel of beautiful rose-coloured flowers, having at its base 3 lanceolate bracts. By the Exe, near Exeter. Alpington. Topsham marshes. (E. B. t. 651.) P. VI. VII.

Ord. XCIX. ALISMACEÆ.

ALISMA. WATER-PLANTAIN.

1. **A. Plantago** (*greater Water-Plantain.*)—By the edges of ditches, ponds, and lakes. Leaves all radical, oval heart-shaped or lanceolate, on long stalks, large. Flower-stalks 2 or 3 feet high, panicled with whorled branches; flowers small, pale rose-colour. Side of a ditch near the old barn at Torre Abbey. Forde bog. Bovey Heath, etc. (E. B. t. 837.) P. VII. VIII.

2. **A. ranunculoides** (*lesser W.*)—In wet ditches and turfy bogs. Leaves radical, narrow-lanceolate or linear, on long stalks. Whole plant much smaller than the last, the flowers however are larger and paler, the flower-stalks umbellate; the fruit of this is globose, while that of *A. Plantago* is bluntly triangular. Preston, near Kingsteignton, *Fl. D.* Boggy ground at Lindridge. (E. B. t. 326.) P. VI. VII.

SAGITTARIA. ARROW-HEAD.

S. sagittifolia (*common Arrow-head.*)—In ditches and rivers. Leaves all radical, on very long stalks, rising to the surface of the water, truly arrow-shaped, with lanceolate straight lobes. Flower-stalk upright, leafless, longer than the leaves, the upper part bearing several whorls of handsome white flowers, the upper flowers being staminiferous, and the lower pistilliferous. River Clyst, near Bishop's Clyst Bridge, *Fl. D.* (E. B. t. 84.) P. VII. VIII.

Ord. C. JUNCAGINACEÆ.

TRIGLOCHIN. ARROW-GRASS.

1. **T. palustre** (*marsh Arrow-grass.*)—Wet and marshy places. Leaves all radical, linear, succulent, from 2 or 3 to 8 inches in length. Flower-stalk 8 or 10 inches high, bearing a loose, simple spike of small yellowish-green flowers. Fruit of 3 combined capsules, linear. Petit Tor. Watcombe. Exmouth. (E. B. t. 366.) P. VI. VII.

2. **T. maritimum** (*seaside A.*)—In salt-marshes. Larger and stouter than the last; leaves more succulent; flowers very similar. Fruit ovate, formed of 6 combined capsules. Good-

rington Marsh. Topsham and Exminster marshes. (E. B. t. 255.) P. VII. VIII.

Ord. CI. TYPHACEÆ.

TYPHA. CAT'S-TAIL, OR REED-MACE.

T. latifolia (*great Reed-Mace.*)—Margins of ponds and lakes. Stems from 3 to 6 feet high; leaves very long, and sometimes nearly an inch broad, erect and linear. Flowers terminating the stem, in two continuous cylindrical spikes; the upper male spike yellow; the lower female one dark brown and velvety. When in fruit the upper spike becomes a bare stalk, while the lower one enlarges considerably from the swelling of the seeds. Hackney clay-pits, near Kingsteignton. Near Chudleigh. (E. B. t. 1455). P. VII. VIII.

SPARGANIUM. BUR-REED.

1. **S. ramosum** (*branched Bur-Reed.*)—By the sides of ditches, lakes, and ponds. Root-leaves very long, linear, sword-shaped, triangular at the base; stems 2 feet or more high, with a few long linear leaves, having broad membranous sheaths at their base. Flowers in a scattered panicle, formed of from 3 to 6 or 8 simple branches, each bearing several heads of male and female flowers. Ditches near Torre Abbey. Paignton. (E. B. t. 744.) P. VII.

2. **S. simplex** (*unbranched upright B.*)—In ditches and by stagnant pools. Smaller than the last; leaves narrower. Stem unbranched; flowers sessile at the summit of the simple stem, the lowest female flower however having a short flower-stalk. River Exe, below Cowley Bridge. Clyst, near Clyst Bridge, *Fl. D.* (E. B. t. 745.) P. VII.

Ord. CII. ARACEÆ.

ARUM. CUCKOO-PINT.

A. maculatum (*Cuckoo-pint, or Wake-robin.*)—In copses and woods, under hedges, etc. Leaves large, radical, on long stalks, ovate-hastate, of a shining green, frequently spotted with black or purple. Spatha, or sheath in which the spike is contained, 6 or 8 inches long, swelling out above and below, constricted in the centre, and tapering to a point at the top, the edges turned in when open. Spike mostly concealed by the spatha, the yellow or purple club-shaped top alone appearing; beneath the club is a row of filaments, then whorls of sessile anthers, and at the base several circles of ovaries which, after the flowering, ripen into scarlet berries, and remain after the upper part of the plant has disappeared. Very abundant. (E. B. t. 1298.) P. IV. V.

ORD. CIII. ORONTIACEÆ.

ACORUS. SWEET-SEDGE.

A. Calamus (*common Sweet-Sedge.*)—On the banks of lakes and streams. Leaves 2 or 3 feet long, linear, erect. Flower-stem with a long leaf-like prolongation beyond the spike. Spike 2 or 3 inches long, lateral, covered completely by the yellowish-green flowers. Whole plant when crushed sweet-scented. New Cut, near Exeter, Haine banks, Exeter, *Fl. D.* Most probably introduced. (E. B. t. 536.) P. VI.

ORD. CIV. PISTIACEÆ.

LEMNA. DUCKWEED.

1. **L. trisulca** (*ivy-leaved Duckweed.*)—On ponds and clear stagnant waters. Fronds (or leaves) floating on the water, about half an inch long, elliptic-lanceolate, minutely toothed at one end and tailed at the other, having 2 young fronds springing from opposite sides, and a single thread-like root from beneath. Flowers proceeding from a fissure in the edge of the frond. Exminster, *Mr. Earle.* (E. B. t. 926.) A. VI.

2. **L. minor** (*lesser D.*)—On ponds and stagnant water. Fronds about ⅛ of an inch long, roundish or broadly ovate, attached together in patches and floating on the surface, with one thread-like root under each. Flowers like the last. Very common. (E. B. t. 1095.) A. VI.

3. **L. polyrrhiza** (*greater D.*)—Still waters. Fronds larger than in the other species, about ½ an inch long and the same in breadth, nearly orbicular, convex below; the under surface and the margin tinged with purple; roots from each frond numerous. Flowers never seen in England. Ponds at Teigngrace. Ilsington. Topsham. (E. B. t. 2458.) A.

4. **L. gibba** (*gibbous D.*)—Stagnant waters, rare. Fronds similar to but larger than *L. minor*, thick, flat on the upper but nearly hemispherical and cellular on the under side; a single root descends from each frond. Flower with 2 stamens. Between Bishopsteignton and Kingsteignton. (E. B. t. 1233.) A. VI.–IX.

Ord. CV. NAIADACEÆ.

POTAMOGETON. PONDWEED.

1. **P. densus** (*opposite-leaved Pondweed.*) — In ditches. Leaves submersed, thickly crowded together, opposite, pellucid, clasping the stem, pointed-ovate or lanceolate. Spikes stalked, 4-flowered. Flowers green. River Clyst, by St. Mary Clyst Bridge, *Fl. D.* (E. B. t. 397.) P. VI. VII.

2. **P. pectinatus** (*fennel-leaved P.*)—In rivers, ponds, and salt-marshes. Leaves alternate, submersed, very narrow and pointed, sheathing at the base, growing from the slender stems in opposite directions. Flower-spikes rising above the water. Flowers of an olive-green, in interrupted clusters. Goodrington Marsh. (E. B. t. 323.) P. VI. VII.

3. **P. pusillus** (*small P.*)—In ditches and stagnant waters. Leaves submersed, alternate, narrow-linear, sessile. Flower-stalks bearing a spike of 3 or 4 greenish flowers. The var. β, *P. compressus*, E. B. t. 418, with broader leaves and attaining a greater size in all its parts, is united with this by Hooker and Arnott. Ditches near Powderham. Exminster marshes. (E. B. t. 215.) P. VI. VII.

4. **P. crispus** (*curly P.*)—In ditches and streams. Leaves alternate, pellucid, linear-oblong, blunt, waved, and serrated, sessile. Flower-stalks longer than the leaves, bearing a lax spike of purplish-brown flowers. Mill-stream near Totness. River Exe. (E. B. t. 1012.) P. VI.

5. **P. perfoliatus** (*perfoliate P.*)—In lakes, ditches, and streams. Stem long and slightly branched. Leaves all submersed, alternate, ovate heart-shaped, clasping the stem, pellucid. Flower-spikes short; flowers dullish purple. River Exe, between Topsham and Exeter, *Fl. D.* (E. B. t. 168.) P. VII.

6. **P. lucens** (*shining P.*)—In lakes, pools, and streams. Stems long, round, and leafy. Leaves submersed, alternate, pellucid, stalked, ovate or lanceolate, accompanied by a winged stipule. Spike dense and many-flowered. Flowers dark green. Pools on Haldon. (E. B. t. 376.) P. VI.

7. P. rufescens (*reddish P.*)—In ditches and slow streams. Submersed leaves lanceolate, narrowed at both ends, pellucid; floating leaves obovate, blunt, narrowed into a short stalk, somewhat leathery in texture, having usually a purplish tinge. Spikes on long stalks, dense. Flowers reddish. Forde bog. (E. B. t. 1286.) P. VII.

8. P. natans (*sharp-fruited broad-leaved P.*)—In ditches, ponds, and slow streams. Stems round and much branched. Lower submersed leaves linear; upper leaves numerous, floating, on long stalks, ovate, 2 or 3 inches long. Flower-stalks axillary, round, contracted just beneath the spike. Flowers dull green. Forde bog. Bovey Heath. (E. B. t. 1822.) P. VI. VII.

RUPPIA. RUPPIA.

R. maritima (*sea Ruppia.*)—In salt-water marshes and ditches. Stems long and slender, much branched. Leaves alternate, linear, with an inflated membranous sheath at their base. Flower-stalk long and spirally twisted. Nut ovoid. Goodrington Marsh. Marshes at Exminster and Powderham. Salt-ditches near Starcross. (E. B. t. 136.) B. VII. VIII.

ZANICHELLIA. HORNED PONDWEED.

Z. palustris (*common horned Pondweed.*)—In ditches and stagnant water. Plant floating. Stems long, thread-like, and branched. Leaves opposite, linear, having a grassy appearance. Flowers axillary, sessile; sterile flower with one stamen, bearing a 4-celled anther; fertile one with 4 or 5 stalked germens. Mill-pond at Lympstone, *Fl. D.* (E. B. t. 1844.) A. or P. V.-VIII.

ZOSTERA. GRASS-WRACK.

Z. marina (*broad-leaved Grass-wrack.*)—In the sea, from whence it is thrown up by the tide. Leaves grassy, long and linear. Flowers imperfect, the stamens and pistils inserted in two rows upon one side of the flat, thin spadix. Frequent on the shores of Torbay. Teignmoth. Exmouth, etc. (E. B. t. 467.) P. ? VII. VIII.

SUB-CLASS III. **GLUMACEÆ.** (ORD. CVI. CVII.)

ORD. CVI. **CYPERACEÆ.**

As the plants belonging to this and the following Order cannot be understood without a more minute description of details than it enters into the design of this work to give, I shall merely set before my readers a catalogue of the names and the several habitats of such of them as grow in this neighbourhood; referring those persons who wish to study them, to the excellent Manuals of the British Flora mentioned in my Preface.

SCHŒNUS. BOG-RUSH.

S. nigricans (*black Bog-Rush.*)—In wet moors and boggy places. Forde bog. Bovey Heath. Chudleigh. Boggy heath at Lindridge, near Bishopsteignton. (E. B. t. 1121.) P. V.–VII.

RHYNCHOSPORA. BEAK-RUSH.

R. alba (*white beak-Rush.*)—In turfy bogs and wet pastures. Forde bog. Bovey Heath. Bogs in Dartmoor. *Schœnus*, Linn. (E. B. t. 985.) P. VI.

ELEOCHARIS. SPIKE-RUSH.

1. **E. palustris** (*creeping Spike-Rush.*)—In wet and marshy places. Wet meadows at Paignton. Forde bog. Bovey Heath. Boggy places in Dartmoor. (*Scirpus*, E. B. t. 131.) P. VI.

2. **E. multicaulis** (*many-stalked S.*)—In marshy places. Near the coal-pits at Bovey Heathfield. (*Scirpus*, E. B. t. 1187.) P. VII.

ISOLEPIS. MUD-RUSH.

1. **I. fluitans** (*floating Mud-Rush.*)—In ditches and ponds. Bovey Heath. Knighton Heath, near Chudleigh. *Scirpus*, Linn. (E. B. t. 216.) P. VI. VII.

2. **I. setacea** (*bristle-stalked M.*)—In wet sandy and gravelly

places, frequent. Forde bog. (*Scirpus setaceus*, E. B. t. 1693.) P.? VII. VIII.

SCIRPUS. CLUB-RUSH, BULRUSH.

1. **S. lacustris** (*lake Club-Rush, or Bulrush.*)—On the margins of ponds and lakes. Goodrington Marsh. Pond between Newton and Chudleigh. Banks of the Exe. (E. B. t. 666.) P. VI. VII.

2. **S. Tabernæmontani** (*glaucous C.*)—In rivers and ponds. Starcross. Ditches at Powderham. *S. glaucus*, Sm. (E. B. t. 2321.) P. VI. VII.

3. **S. maritimus** (*salt-marsh C.*)—Frequent in salt-marshes. Goodrington Marsh. Exminster marshes. (E. B. t. 542.) P. VII.

4. **S. sylvaticus** (*wood C.*)—In damp woods and sides of rivers. Banks of the Teign near Chudleigh Bridge. Kingsteignton, *Fl. D.*

5. **S. cæspitosus** (*scaly-stalked C.*)—In moist heathy places and on moors, very common. Forde bog. Milber Down, etc. (E. B. t. 1029.) P. VI.-VIII.

ERIOPHORUM. COTTON-GRASS.

1. **E. vaginatum** (*hair-tail Cotton-Grass.*) — In bogs and moors. Forde bog. Haldon. Boggy places in Dartmoor. (E. B. t. 873.) P. V.

2. **E. latifolium** (*broad-leaved C.*)—In bogs, marshes, and heaths. Not frequent. Bovey Heath and Bovey Tracey, *Mr. C. E. Parker*. (*E. polystachion*, E. B. t. 563.) P. V. VI.

3. **E. angustifolium** (*narrow-leaved C.*)—In turf-bogs, meadows, and moors, common. Babbicombe. Petit Tor. Forde bog, etc. (E. B. t. 564.) P. V. VI.

KOBRESIA. KOBRESIA.

K. caricina (*compound-headed Kobresia.*) — Moors and heaths. Haldon Down, plentiful, *Fl. D. Schœnus monoicus*, Sm. (E. B. t. 1410.) P. VII.

CAREX. CAREX, OR SEDGE.

1. **C. pulicaris** (*flea Carex.*)—Frequent in boggy places. Forde bog. Wet heath at Lindridge. Knighton Heath, near Chudleigh. Haldon, etc. (E. B. t. 543.) P. v. vi.

2. **C. ovalis** (*oval spiked C.*)—In bogs and marshy spots. Knighton Heath, near Chudleigh. Goodrington Marsh. Forde bog. (E. B. t. 306.) P. vi.

3. **C. stellulata** (*little prickly C.*)—In marshy and heathy places. Knighton Heath, near Chudleigh. Bovey Heath, etc. (E. B. t. 806.) P. v. vi.

4. **C. curta** (*white C.*)—In bogs and wet places, but not very frequent. Wet places about Chudleigh, *Fl. D.* (E. B. t. 1386.) P. vi.

5. **C. remota** (*distant-spiked C.*)—In woods and damp shady places, frequent.' Wood near Ansti's Cove. Ilsham, etc. (E. B. t. 832.) P. v. vi.

6. **C. axillaris** (*axillary-clustered C.*)—In marshes and wet situations. Watery spot in the thicket near Ilsham Farm. Marychurch. Forde bog. (E. B. t. 993.) P. v. vi.

7. **C. paniculata** (*great panicled C.*)—In bogs, and wet, swampy situations. Forde bog. Knighton Heath, near Chudleigh. Topsham causeway. (E. B. t. 1064.) P. v. vi.

8. **C. teretiuscula** (*lesser panicled C.*)—In bogs and watery meadows. Forde bog. Boggy heath at Lindridge. Ilsington. (E. B. t. 1065.) P. vi.

9. **C. vulpina** (*great C.*)—In wet, boggy situations, and on banks of rivers. Forde bog. Banks of Teign, in many situations. Chudleigh. (E. B. t. 307.) P. vi.

10. **C. divulsa** (*grey C.*)—In damp shady situations. Chudleigh, *Mr. C. Parker.* (E. B. t. 629.) P. v. vi.

11. **C. muricata** (*greater prickly C.*)—In damp gravelly pastures and under hedges. Berry Pomeroy woods. Bradley woods. Chudleigh. (E. B. t. 1097.) P. v. vi.

12. **C. arenaria** (*sea C.*)—On sandy seashores, very frequent. Paignton sands. Goodrington, etc. (E. B. t. 928.) P. v. vi.

13. **C. intermedia** (*soft brown C.*)—In wet ground and marshy meadows. Meadows near Torre Abbey. Wet places about Chudleigh. Bovey Heath. (E. B. t. 2042.) P. vi.

14. **C. vulgaris** (*common C.*)—In marshy places and wet pastures, very frequent. Banks above Meadfoot. Ilsham. Forde bog. Paignton, etc. (E. B. t. 1507.) P. v. vi.

15. **C. acuta** (*slender-spiked C.*)—In moist meadows and pastures, frequent. Forde bog. Bovey Heath. Paignton. (E. B. t. 580.) P. v.

16. **C. extensa** (*long-bracteated C.*)—Seaside marshes, rare. Exmouth, *Fl. D.* (E. B. t. 833.) P. vi.

17. **C. flava** (*yellow C.*)—Marshy places and turfy bogs, common. Paignton meadows. Forde bog. Heath at Lindridge, etc. (E. B. t. 1294.) *C. Œderi*, E. B. t. 1773, is found growing in the same habitats. P. v. vi.

18. **C. distans** (*loose C.*)—Marshy places, generally near the sea. Forde bog. Bovey Heath. Knighton Heath, near Chudleigh. Haldon. (E. B. t. 1234.) P. vi.

19. **C. binervis** (*green-ribbed C.*)—To be found in the same localities as the last. (E. B. t. 1235.) P. vi.

20. **C. depauperata** (*starved wood C.*)—In dry woods and hedges. Berry Pomeroy woods. Bradley woods. (E. B. t. 1098.) P. v. vi.

21. **C. panicea** (*pink-leaved C.*)—In bogs and marshy places, frequent. Paignton. Forde bog. Meadows at Newton, near the Teign. (E. B. t. 1505.) P. v. vi.

22. **C. pallescens** (*pale C.*)—In wet and marshy places, frequent. Same situations as the last. (E. B. t. 2185.) A. v. vi.

23. **C. sylvatica** (*pendulous wood C.*)—In moist and shady woods, frequent. Thicket between Ilsham Farm and Ansti's Cove lane. Berry Pomeroy woods, etc. (E. B. t. 995.) P. v. vi.

24. **C. Pseudo-Cyperus** (*Cyperus-like C.*)—In damp places, by the sides of rivers, ponds, and lakes. Stover canal. Banks of canal, near Exeter. St. John's-in-the-Wilderness, near Exmouth, *Fl. D.* (E. B. t. 242.) P. vi.

25. **C. pendula** (*great pendulous C.*)—In damp woody and shady places. Meadows near Moreton. Near Exmouth, *Fl. D.* (E. B. t. 2315.) P. v. vi.

26. **C. glauca** (*glaucous heath C.*)—In wet pastures, moors, and heaths. Forde bog. Wet heath at Lindridge, etc. (*C. recurva*, E. B. t. 1506.) P. vi.

27. **C. digitata** (*fingered C.*)—In woods, etc., in limestone countries, rare. Copse close by Forde bog. (E. B. t. 615.) P. v.

28. **C. præcox** (*vernal C.*)—In dry pastures and heaths, very frequent. Cliffs at Meadfoot. Walks near Hope's Nose. Paignton. Forde bog, etc. (E. B. t. 1099.) P. iv. v.

29. **C. pilulifera** (*round-headed C.*)—In bogs and moors,

very frequent. Forde bog. Bovey Heath, etc. (E. B. t. 885.) P. VI.

30. **C. hirta** (*hairy C.*)—In damp woods and wet pastures, frequent. Chudleigh wood. Berry Pomeroy woods, etc. (E. B. t. 685.) P. V. VI.

31. **C. vesicaria** (*short-beaked bladder C.*)—In bogs and marshes. Knighton Heath, near Chudleigh. Near Topsham. (E. B. t. 779.) P. V. VI.

32. **C. paludosa** (*lesser common C.*)—In ditches and on banks of rivers, common. Paignton. Forde bog, etc. (E. B. t. 807.)

33. **C. riparia** (*great common C.*)—By the sides of ditches and streams. Banks of the water in front of Forde House. Topsham marshes. Exminster Marsh, etc. (E. B. t. 579.) P. IV. V.

Ord. CVII. **GRAMINEÆ.**

Those who wish to distinguish the different genera of this Order will be greatly assisted by consulting plates 6-9, at the end of Hooker and Arnott's 'British Flora;' Parnell's 'Grasses of Britain,' containing excellent figures of the numerous species, and Sowerby's plates of 'Grasses' in 'English Botany,' may also be studied with considerable advantage.

ANTHOXANTHUM. VERNAL-GRASS.

A. odoratum (*sweet-scented Vernal-Grass.*)—In meadows, woods, and pastures, very abundant, Walks at Meadfoot. Meadows about Torquay, etc. (E. B. t. 647.) P. V. VI.

NARDUS. MAT-GRASS.

N. stricta (*Mat-Grass.*)—Most abundant on all our heaths and moors, constituting often the chief part of the turf. (E. B. t. 290.) P. VI.

ALOPECURUS. FOXTAIL-GRASS.

1. **A. pratensis** (*meadow Foxtail-Grass.*)—Common in meadows and pastures. Fields, etc., about Torquay, Marychurch, Paignton, etc. (E. B. t. 759.) P. V. VI.

2. **A. agrestis** (*slender F.*)—In fields and by waysides. Under hedges and by roadsides in the neighbourhood of Torquay. Common. (E. B. t. 848.) A. v.-vii.

3. **A. bulbosus** (*tuberous F.*)—In salt-marshes, rare. Goodrington Marsh. (E. B. t. 1249.) P. v.-vii.

4. **A. geniculatus** (*floating F.*)—In pools and wet and marshy places. Goodrington Marsh. (E. B. t. 1250.) P. v.-viii.

PHALARIS. CANARY-GRASS.

1. **P. Canariensis** (*cultivated Canary-Grass.*)—Found sometimes in fields and by roadsides, but not indigenous. Side of the cliffs by the New Road, *Mr. C. E. Parker.*

2. **P. arundinacea** (*reed C.*)—Banks of rivers and lakes. Stream behind Forde House. Banks of Teign, in many places. (E. B. t. 402.) P. vii. viii.

PHLEUM. CAT'S-TAIL GRASS.

1. **P. pratense** (*common C., or Timothy-Grass.*) Meadows and pastures. About Torquay, etc., very common. (E. B. t. 1076.) P. vi.-x. The variety *P. nodosum,* in fields at Marychurch.

2. **P. arenarium** (*sea C.*)—In sandy situations near the sea. Paignton sands. (*Phalaris,* E. B. t. 222.) A. v. vi.

MILIUM. MILLET-GRASS.

M. effusum (*spreading Millet-Grass.*)—In damp and shady woods. Bradley woods. (E. B. t. 1106.) P. v. vi.

GASTRIDIUM. NIT-GRASS.

G. lendigerum (*awned Nit-Grass.*)—Damp places near the sea, rare. Babbicombe, slope leading from the down to the beach. (*Milium,* E. B. t. 1107.) P. vi.-ix.

CALAMAGROSTIS. SMALL-REED.

C. lanceolata (*purple-flowered Small-Reed.*)—In wet places. Ditches near Torre Abbey. Paignton. *Arundo Calamagrostis*, Linn. (E. B. t. 2159.) P. VII.

AMMOPHILA. SEA-REED.

A. arundinacea (*common Sea-Reed, Marum, or Matweed.*)—On sandy seashores. Teignmouth. Exmouth sands. (*Arundo arenaria*, E. B. t. 520.) *Psamma*, Bab. P. VII.

AGROSTIS. BENT-GRASS.

1. **A. canina** (*brown Bent-Grass.*)—In moist heaths and moors. Forde bog. Bovey Heath, etc. (E. B. t. 1856.) P. VI. VII.

2. **A. setacea** (*bristle-leaved B.*)—On dry downs, but very local, confined to the south and south-west parts of England. Milber Down. Kerswell Down. Holne Chase. (E. B. t. 2138.) P. VI. VII.

3. **A. vulgaris** (*fine B.*)—In meadows, pastures, and on hedge-banks. Common everywhere about the neighbourhood. (E. B. t. 1671.) P. VII.

4. **A. alba** (*marsh B.*)—In pastures, meadows, and by roadsides. Fields about Torquay and the vicinity. (E. B. t. 1189.) P. VI. VIII. The variety β, *A. stolonifera*, is found at Torquay and on the rocks around Torbay.

AIRA. HAIR-GRASS.

1. **A. cæspitosa** (*tufted Hair-Grass.*)—In damp shady places, and borders of meadows. Base of the cliffs on the new sea-road, near the turnpike. (E. B. t. 1453.) P. VI. VII.

2. **A. flexuosa** (*waved H.*)—Abundant on heaths and hilly pastures. Meadfoot cliffs. Daddyhole Plain, on the descent to the quarry, etc. (E. B. t. 1519.) P. VII.

3. **A. caryophyllea** (*silvery H.*)—On hills and pastures, fre-

quent. Meadfoot, base of the cliffs by the sea-road. Daddyhole Plain. Kerswell Down, etc. (E. B. t. 812.) P. v. vii.

4. **A. præcox** (*early H.*)—On hills and dry pastures, common, Meadfoot. Paignton sands, etc. (E. B. t. 1296.) A. v. vi.

MOLINIA. MOLINIA.

M. cærulea (*purple Molinia.*)—On wet heaths and moors. Milber Down. Bovey Heath. Lindridge Heath. Haldon. Various parts of Dartmoor. (*Melica*, E. B. t. 750.) P. vii. viii.

MELICA. MELIC-GRASS.

M. uniflora (*wood M.*)—Frequent in shady woods and thickets. Bradley woods. Berry Pomeroy woods. Chudleigh, etc. (E. B. t. 1058.) P. v.–vii.

HOLCUS. SOFT-GRASS.

1. **H. mollis** (*creeping Soft-Grass.*)—Frequent in all our meadows, pastures, and hedgesides. (E. B. t. 1170.) P. vii.

2. **H. lanatus** (*meadow S.*)—In meadows, pastures, and woods. Very frequent. (E. B. t. 1169.) P. vi. vii.

ARRHENATHERUM. OAT-LIKE GRASS.

A. avenaceum (*common oat-like Grass.*)—Frequent in hedges and pastures. This and the var. *A. nodosum* are found in meadows about Torquay and on the cliffs at Meadfoot. (*Holcus avenaceus*, E. B. t. 813.) P. vi. vii.

KŒLERIA. KŒLERIA.

K. cristata (*crested Kœleria.*)—In dry pastures. Daddyhole Plain, side of the descent to the quarry. Marychurch. *Aira*, Linn. (E. B. t. 648.) P. vi. vii.

POA. MEADOW-GRASS.

1. **P. aquatica** (*Reed Meadow-Grass.*)—By the sides of ponds, rivers, and ditches. Side of the stream in Torre Abbey meadow. Pool of water near Bishopstowe, opposite the path leading on to Babbicombe Down. Goodrington Marsh. Meadow behind Forde House. (E. B. t. 1315.) *Glyceria*, Bab. P. VI. VII.

2. **P. fluitans** (*floating M., or Flote-Grass.*)—Common in ditches and watery places. Paignton meadows. Marychurch, etc. (E. B. t. 1520.) *Glyceria*, Bab. P. VI.-IX.

3. **P. maritima** (*creeping sea M.*)—In damp places by the seacoast. Paignton sands. Teignmouth. Exmouth. (E. B. t. 1140.) *Sclerochloa*, Bab. P. VI. VII.

4. **P. distans** (*reflexed M.*)—Waste sandy places and near the seashore. Hennock, *Mr. C. E. Parker*. Sandy spots about Exmouth, *Fl. D.* (E. B. t. 986.) *Sclerochloa*, Bab. P. VI.-VIII.

5. **P. procumbens** (*procumbent sea M.*)—In salt-marshes. Paignton. Goodrington Marsh. (E. B. t. 532.) *Sclerochloa*, Bab. A. VI. VII.

6. **P. rigida** (*hard M.*)—On old walls, rocks, and in dry barren ground. Walls about Torquay. Daddyhole Plain, on the descent to the quarry. Kerswell Down. Milber Down. Rocky places about Dartmoor, etc. (E. B. t. 1371.) *Sclerochloa*, Bab. A. V. VI.

7. **P. compressa** (*flat-stemmed M.*)—On tops of walls and in dry situations. Old walls at Torre, *Mr. C. E. Parker*. Rocks at Chudleigh. Nutwell, near Lympstone, *Fl. D.* (E. B. t. 365.) P. VI. VII. β. *P. polynoda*, Parnell's Grasses, tt. 39, 91, 92. Down above Ansti's Cove, *Mr. C. E. Parker*.

8. **P. pratensis** (*smooth-stalked M.*)—Frequent in meadows and pastures. Growing abundantly about Torquay and the neighbourhood. (E. B. t. 1073.) P. VI.-VII.

9. **P. trivialis** (*roughish M.*)—Common in meadows and pastures. Under hedges and in meadows, etc., about Torquay and Marychurch. (E. B. t. 1072.) A. VI. VII.

10. **P. bulbosa** (*bulbous M.*)—On sandy seashores. Paignton sands. Teignmouth Den. (E. B. t. 1071.) A. IV. V.

11. **P. annua** (*annual M.*)—In meadows and pastures, under hedges, and by roadsides. Very abundant everywhere. (E. B. t. 1141.) A. III.-IX.

TRIODIA. HEATH-GRASS.

T. decumbens (*decumbent Heath-Grass.*)—On heaths, moors, and dry mountain pastures. Haldon, Hightor Down, *Fl. D.* Middleton Down, near Chagford. Should be looked for on Milber Down. (*Poa*, E. B. t. 792.) P. VII.

BRIZA. QUAKING-GRASS.

B. media (*common Quaking-Grass.*)—In meadows and pastures, growing abundantly. Meadfoot. Daddyhole Plain. Meadows on the Warberry Hill, etc. (E. B. t. 340.) P. VI.

DACTYLIS. COCK'S-FOOT GRASS.

D. glomerata (*rough cock's-foot Grass.*)—By roadsides, in meadows and woods. Very abundant everywhere in the neighbourhood of Torquay. (E. B. t. 335.) P. VI. VII.

CYNOSURUS. DOG'S-TAIL GRASS.

C. cristatus (*crested dog's-tail Grass.*)—In dry fields and pastures. Torre Abbey meadows. Side of Paignton road. Chelston meadows. Meadow by Livermead Cottage. Marychurch, abundant. (E. B. t. 316.) P. VII.

FESTUCA. FESCUE-GRASS.

1. **F. uniglumis** (*single-glumed Fescue-Grass.*)—On sandy seashores. Base of the cliffs at Meadfoot. Exmouth sands. (E. B. t. 1430.) A. VI.

2. **F. bromoides** (*barren F.*)—On old walls and in dry sandy places. Park Hill. Daddyhole Plain. Paignton Green. Kerswell Down, etc. (E. B. t. 1411.) *F. sciuroides*, Bab. A. ? VI. VII.

3. **F. myurus** (*wall Fescue-Grass.*)—In similar situations to the last. Kingskerswell, on old walls. Near Forde bog, *Mr. C. E. Parker.* Waste ground by the new cut at Meadfoot. (E. B. t. 1412.) A. ? VI. VII.

4. **F. ovina** (*sheep's F.*)—On dry hilly pastures. Daddyhole Plain, on the descent to the quarry. Ilsham, etc. (E. B. t. 585.) Var. β, *F. duriuscula*, E. B. t. 470. Ilsham, Marychurch. P. VI.

5. **F. pratensis** (*meadow F.*)—In damp meadows and pastures, common. Side of meadow near the old gate-house on the Babbicombe road. Fields at Marychurch, etc. (E. B. t. 1592.) The var. β (if it is not a really distinct species), *F. loliacea*, I have found on Meadfoot cliffs, Daddyhole Plain, on the slope to the quarry, and on Kerswell Down (E. B. t. 1821), much coarser in all its parts than my specimens. P. VI. VII.

6. **F. elatior** (*tall F.*)—In damp pastures and sides of rivers. Waldon Hill (formerly: habitat destroyed by building). Meadows about Torquay. Paignton meadows. Shaldon. Teignmouth. (E. B. t. 1593.) *F. arundinacea*, Bab., grows in great plenty at the base of the Meadfoot cliffs by the side of the sea-road, and sparingly by the side of new road under the Waldon Hill, near the turnpike. (P. VI. VII.)

7. **F. gigantea** (*tall bearded F.*)—In moist woods and thickets. Copse near Ansti's Cove. Wood by the Newton road. Woods at Shiphay. (E. B. t. 1820.) P. VII.

BROMUS. BROME-GRASS.

1. **B. erectus** (*upright Brome-Grass.*)—In dry sandy fields and by roadsides. Chudleigh, *Mr. C. E. Parker.* (E. B. t. 471.) P. VI. VII.

2. **B. asper** (*hairy wood B.*)—Under hedges, and in damp woods and thickets. Frequent by the hedges in lanes about Cockington and Shiphay. Marychurch. Berry Pomeroy woods. (E. B. t. 1172.) A. or B. VII.

3. **B. sterilis** (*barren B.*)—In waste places. Very abundant everywhere. Rock Walk. Meadfoot cliffs. Hope's Nose. Sides of nearly all the roads and lanes. (E. B. t. 1030.) A. VI.

4. **B. diandrus** (*upright annual B.*)—In dry sandy places. Not common. Rocky valley to the right of the Teignmouth road, *Mr. C. E. Parker.* (E. B. t. 1006.) A. VI. VII.

5. **B. secalinus** (*smooth Rye B.*)—In cornfields, occasionally. Fields near Hope's Nose, *Mr. C. E. Parker.* (E. B. t. 1171.) *Serrafalcus*, Bab. *B. velutinus*, Sm. (E. B. t. 1884), is also to be found in the same locality. A. or B. VI. VII.

6. **B. commutatus** (*tumid field B.*)—Common by roadsides and in cornfields. Barton. Marychurch, etc. (*B. pra-*

150   GLUMACEÆ.

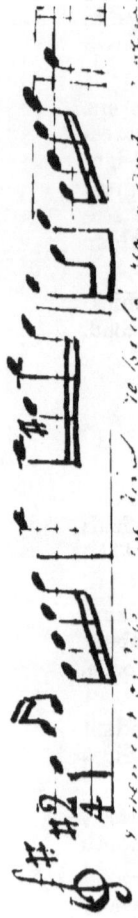

tensis, E. B. t. 920.) *Serrafalcus*, Bab. The variety *multiflorus* of Parnell was found at Ilsham by Mr. C. E. Parker. B. VI. VII.

7. **B. racemosus** (*smooth B.*)—In meadows and pastures. Fields at Chelston, Cockington, etc. (E. B. t. 1079.) *Serrafalcus*, Bab. B. VI.

8. **B. mollis** (*soft B.*)—In meadows and pastures, on banks, and by roadsides; frequent everywhere. The varieties *ovalis* and *nanus* of Parnell grow on Paignton sands. (E. B. t. 1078.) *Serrafalcus*, Bab. A. or B. V. VI.

9. **B. arvensis** (*taper field B.*)—In fields, occasionally. Kingsteignton, Bishopsteignton, *Fl. D.* According to Mr. Babington, "not even naturalized." (E. B. t. 1984.) A. VII. VIII.

AVENA. OAT, OAT-GRASS.

1. **A. fatua** (*wild Oat.*)—In cornfields, occasionally. Fields on the Warberry Hill. Ilsham. Marychurch. (E. B. t. 2221.) A. VII.

2. **A. strigosa** (*bristle-pointed O.*)—Common in cornfields. Fields at Torquay, Ilsham, and Marychurch. (E. B. t. 1266.) A. VII.

3. **A. pratensis** (*narrow-leaveed perennial O.*)—In dry pastures and heathy and mountainous situations. Near Daddyhole Plain. Babbicombe. Newton. Chudleigh. (E. B. t. 1204.) P. VI.

4. **A. pubescens** (*downy O.*)—In dry pastures, in chalky and limestone districts. Ilsham Down. Babbicombe. Kerswell Down, etc. (E. B. t. 1640.) P. VII.

5. **A. flavescens** (*yellow O.*)—Frequent in dry meadows and pastures. Walks above Meadfoot. Kingsteignton. Meadows between Bishopsteignton and Lindridge. Chudleigh. (E. B. t. 952.) *Trisetum*, Bab. P. VII.

PHRAGMITES. REED.

P. communis (*common Reed.*)—In ditches, by margins of lakes and rivers, very abundant. Meadow by the side of Torre Abbey. Wet meadows at Paignton. Maidencombe. Forde bog. Teignmouth. Exmouth, etc. *Arundo Phragmites*, Linn. (E. B. t. 401.) P. VII. VIII.

GLUMACEÆ.

ELYMUS. LYME-GRASS.

E. arenarius (*upright sea Lyme-Grass.*)—On sandy seashores. Paignton sands. Goodrington sands. Exmouth sands, etc. (E. B. t. 1672.) P. VII.

HORDEUM. BARLEY.

1. **H. pratense** (*meadow Barley.*)—In damp meadows and pastures. Side of the Babbicombe road, just opposite Wellswood Park. Meadows at Ilsham. Fields near Newton. Chudleigh, etc. (E. B. t. 409.) A. VI. VII.

2. **H. murinum** (*wall B.*)—In waste places, by sides of walls, etc. In dry rocky places and by waysides about Torquay. Teignmouth. Dawlish. Newton, etc. (E. B. t. 1971.) A. VI. VII.

3. **H. maritimum** (*seaside B.*)—In sandy and dry pastures near the sea. Paignton sands. Goodrington sands. Waste ground by the side of the sea walk at Teignmouth. Exmouth sands. (E. B. t. 1205.) A. VI.

TRITICUM. WHEAT, WHEAT-GRASS.

1. **T. junceum** (*rushy sea Wheat-Grass.*)—On the seashore. Side of the New Road. Paignton sands. Teignmouth. Dawlish. Exmouth. (E. B. t. 814.) P. VII. VIII.

2. **T. laxum** (*flat-leaved sea W.*)—On sandy seashores. New Road, near the cliffs. Paignton sands, near the harbour. P. VI.–VIII.

3. **T. repens** (*creeping W., or Couch-Grass.*)—In fields and waste places. Common everywhere about the neighbourhood. (E. B. t. 909.) P. VII. Var. β, *T. littorale*, Paignton sands, Mr. C. E. Parker.

4. **T. caninum** (*fibrous-rooted W.*)—In woods, under hedges, and on banks. Field near Hope's Nose. Marychurch. Paignton. (E. B. t. 1372.) P. VII.

BRACHYPODIUM. FALSE BROME-GRASS.

1. **B. sylvaticum** (*slender false Brome-Grass.*)—In woods

and hedges. Wood on the Newton road. Bradley woods. Berry Pomeroy woods. Chudleigh rocks and the neighbouring woods, *Fl. D.* (*Bromus sylvaticus*, E. B. t. 729.) P. VII.

2. **B. pinnatum** (*heath false B.*)—Open fields and heathy places, on dry limestone soil. Milber Down. Bovey Heath. Chudleigh, Ingsdon near Ilsington, Bovey Tracey, Ashburton, *Fl. D.* (*Bromus pinnatus*, E. B. t. 730.) P. VII.

LOLIUM. DARNEL, RYE-GRASS.

1. **L. perenne** (*perennial or beardless Rye-Grass.*)—By roadsides, in pastures and waste places, etc., very frequent. Growing most abundantly in every pasture and waste ground in the neighbourhood. (E. B. t. 315.) P. VI.

2. **L. Italicum** (*Italian Rye-Grass.*)—In pastures and waste places, occasionally. Piece of waste ground near the new carriage-drive above Meadfoot cliffs. Field near Ilsham Farm. Side of the hedge by the road leading from the Warberry Hill to the Ellacombe lanes. *L. multiflorum*, H. and A. B. or P. VI.

3. **L. temulentum** (*Darnel.*)—In cultivated fields, not so frequent as *L. perenne*. Fields at Marychurch. Ilsham. Chudleigh. Near Woodbury, *Fl. D.* (E. B. t. 1124.) A. VI.-VIII.

LEPTURUS. HARD-GRASS.

L. incurvatus (*sea Hard-Grass.*)—On the seashore, not common. Berry Head. Exmouth. 'Flora Devoniensis' gives as a habitat for this, "Parsonage stile, Lympstone;" apparently not a very likely locality. (*Rotbœllia*, E. B. t. 760.) A. VII.

End of the Phænogamous or Flowering Plants.

Class III. ACOTYLEDONOUS OR CELLULAR PLANTS.

"Whole plant with a cellular structure (except in the true Ferns, which have tubular vessels among the cells, and hence approach the Second Class). There are no real flowers, nothing that can be considered as stamen or pistil. The seeds, or organs of reproduction, are without any distinct embryo, consequently without any cotyledon. This Class corresponds with the Twenty-fourth (*Cryptogamia*) in the Linnæan system."—*Brit. Flora.*

Sub-Class I. FILICES.

In describing the Ferns I have departed from the arrangement adopted in the 'British Flora,' and have followed that used by Mr. Moore in his excellent 'Handbook of British Ferns,' a work in which all the species and their many varieties are most clearly and accurately described, and which those who are desirous of making themselves acquainted with this beautiful order of plants will do well to study.

Ord. I. POLYPODIACEÆ.

POLYPODIUM. POLYPODY.

"Sori without any indusium, globose or ovoid, superficial or immersed, the receptacles terminal or medial on the free veins. Veins simple or forked, from a central costa; *venules* free."—*Moore's Handbook.*

1. **Polypodium vulgare** (*common Polypody.*)—On walls, shady rocks, banks, decayed stumps of trees, and old thatched roofs. Fronds deeply pinnatifid; lobes linear-oblong, obscurely serrate, becoming gradually smaller towards the apex of the frond. Rhizome, or rootstock, creeping, branched, densely covered with brown scales. Stipes, or stalk, nearly equal in length to the leafy part of the frond, and distinctly jointed at the base with the caudex. Fronds varying in figure from strap-shaped or narrow-oblong to a more or less ovate form. Venation of each lobe consisting of a prominent wavy midvein, branching alternately; the lateral branches are again divided into from 3 to 5 small veins, one of which terminates in a sorus, and the others, which are barren, end in transparent knobs, which form a line near the margin of

the lobes. Fructification on the back of the frond, and confined usually to its upper part; clusters large and circular, without any indusium or covering; sori tawny or orange-coloured. Very abundant. In the chasm at Daddyhole Plain. Cockington lanes. Babbicombe and Ilsham Downs. Maidencombe. Forde bog. Bradley woods, etc. (E. B. t. 1149.) Moore, 'Nature-Printed Ferns,' t. 1. The variety, *P. semilacerum*, Moore, 'Nature-Printed Ferns,' t. 2, "fronds pinnatifid and fertile above, bipinnatifid below; lobules distinct, linear, acute, serrate" (*Moore's Handbook*), is found sometimes on the rocks at Ansti's Cove, and on the ruined walls of Berry Pomeroy Castle. P. VII.-X.

2. **P. Phegopteris** (*mountain Polypody, or Beech Fern.*)— In damp woods, in hilly countries, and in the neighbourhood of waterfalls, the spray of which it appears to delight in. "Fronds pinnate; pinnæ linear-lanceolate, united at the base, pinnatifid with linear blunt lobes; *lowest pair of pinnæ turned downwards and forwards*, the rest upwards, clusters marginal" (*Bab. Manual*). Rhizome dark-coloured, extensively creeping, somewhat scaly, and sending forth black wiry roots. Fronds from 4 to 20 inches long, triangular in outline, borne on a stipes as long as or generally longer than itself. Pinnæ very acute, with the exception of the lowest pair, pointing upwards, deeply pinnatifid, those near the apex becoming gradually entire; lowest pair of pinnæ distinct from the rest and minutely stalked; all the others connected with the rachis by their whole width, the basal segments forming at their junction a cross-like figure. Venation of the lobes formed by a slender tortuous midvein giving off alternate branches which do not generally again divide, but which extend to the margin and bear a small sorus near the end of each. The sori are circular and nearly marginal, without any indusium; spore-cases pale brown. Woods at Holne Chase, near Holne Bridge. Beckey Falls. Parts of Dartmoor. (E. B. t. 2224.) Moore, 'Nature-Printed Ferns,' t. 4; Sowerby's Ferns, t. 2. P. VII.-IX.

POLYSTICHUM. SHIELD-FERN.

1. **P. aculeatum** (*common prickly Shield-Fern.*)—On hedge-banks and in woods. *Fronds rigid*, lanceolate or broadly linear-lanceolate, bipinnate, pinnules obliquely decurrent, acute, the anterior basal pinnule much longer than the rest. Caudex thick and tufted; stipes short, usually about 3 or 4 inches long, thickly covered with broad ovate-lanceolate brown scales; rachis thick and also scaly, scales broad below but becoming linear above; fronds from 1 to 3 or more feet high (some which I gathered,

this year, 1859, measured rather more than 4 feet), rigid, dark green; pinnæ numerous, pinnate or lobed and pinnatifid. The first anterior pinnule at the base of each of the pinnæ is always larger and longer the others, and all stand parallel on each side of the main rachis, giving the upper surface of the frond a peculiar appearance, by which it may readily be recognized, pinnules mostly auricled on the anterior side, the auricles acute and, as well as all the principal divisions, terminating in a sharp spinous process. Venation composed of a flexuous midvein, giving off alternate branches, which are again divided, the upper branch giving off 2 or 3 venules and the lower from 3 to 4. Fructification confined generally to the upper half of the frond. Sori round, covered with an indusium, placed in a line on each side of the midvein; indusium round, attached by its centre. Sporecases dark brown, numerous. Very abundantly distributed. Cliffwalks between Meadfoot and Ansti's Cove. Walks above Meadfoot. Cockington lanes. Shiphay lanes. Paignton road. Growing very bold and large in the Old British Road, by the side of the Paignton road. Bradley woods. Maidencombe. Teignmouth road, etc. (*Aspidium*, E. B. t. 1562, and Hook. and Arn. Br. Fl. Moore, Nat. Print. Ferns, t. 10. Sowerby, Ferns, t. 17.) P. VI.-VIII.

2. **P. angulare** (*angular, or soft prickly Shield-Fern.*)— On shady hedgebanks and sheltered woods. "Fronds lax, herbaceous, lanceolate, bipinnate; pinnules distinct, acute or obtuse, with an obtuse-angled base, attached by a distinct stalk, lobed or serrated, the serratures tipped by soft bristles; sori terminal or subterminal" (*Moore's Handbook*). Caudex thick and scaly; stipes longer than in the last, from 4 to 6 inches long, scaly; scales reddish-brown, chaffy, linear-lanceolate. Fronds from 2 to 4 feet high, lanceolate, bi-tripinnate, lax, green, more or less arched or drooping, numerous, growing in a basket from around the crown; the pinnæ are numerous, linear-lanceolate in shape, and taper towards the apex. The first pinnule usually scarcely longer than the rest, though sometimes it is much lengthened; each of the pinnules has a strong anterior auricle, either acute or blunt, serrated, each serrature tipped with a slender bristle; pinnules attached to the secondary rachis by a distinct but short stalk; the venations consist of a flexuous midvein, which is branched alternately, each branch again dividing into 2, 3, or 4 smaller branches, the anterior one of which bears a sorus near its apex; the auricle also has a branched vein which bears 3 or 4 sori. The fructification covers usually the upper two-thirds of the frond, the sori are small and numerous, covered by a round membranaceous indusium, which is attached by its centre; sporecases brown. Cockington lanes. Shiphay lanes. Lane near

Forde bog. Bradley woods, etc. This Fern is more sparingly distributed than the last. (*Aspidium*, E. B. S. t. 2776, and Hook. and Arnott. Moore, Nat. Print. Ferns, t. 12. Sowerby, Ferns, t. 18.) P. VI.–VIII.

LASTREA. BUCKLER-FERN.

1. **L. Filix-mas** (*male Fern, or common Buckler-Fern.*)— On hedgebanks, and in shady and wooded situations. "Fronds bipinnate, pinnules oblong, obtuse, serrate, sori near the central nerve, stipes and rachis chaffy" (*Brit. Flora*). (*Aspidium*, E.B. t. 1458, and t. 1949. *Aspidium*, Hook. and Arnott. Moore, Nat. Print. Ferns, t. 14. Sowerby, Ferns, t. 9.) Caudex large and tufted, when old much lengthened, scaly, and giving off coarse and dark-coloured roots. Stipes about one-third of the whole frond in length, covered with pale brown chaffy scales, the rachis being sparingly sprinkled with smaller ones. Fronds from 2 or 3 to 6 feet in height, arranged in a beautiful circular form around the crown, smooth and of a deep green colour, paler beneath, broadly lanceolate, bipinnate; the pinnæ alternate or nearly opposite, the lower ones more distant than the upper, which narrow by degrees into the acute apex. Pinnules distinct at the lower part of the pinnæ, the remaining pinnules attached by their whole base, and running more or less into one another, oblong, crenated or serrated at their margins; the midvein is slightly tortuous and alternately branched, the branches being again forkedly divided, the ultimate divisions extending nearly to the margin, one branch reaching the point of each marginal tooth. Sori numerous, roundish kidney-shaped, covered by a firm, roundish, but posteriorly notched indusium, attached by the notch. Sporecases reddish brown, the indusium being of a grey or leaden colour. Very abundantly distributed, and found in most of the habitats above given for *Polystichum aculeatum*. The var. *paleacea*, remarkable for its profuse golden-tinted scales and large and spreading fronds, grows plentifully in Berry Pomeroy woods, and in Sharpham woods, on the banks of the Dart. P. VI.–VIII.

2. **L. cristata**, var. *spinulosa*; "Fronds narrow oblong-lanceolate, bipinnate; pinnules oblong-acute, inciso-serrate or pinnatifid, with aristately-toothed lobes; posterior basal pinnules much larger than the anterior ones" (*Moore's Handbook*). In wet and boggy situations. (*Aspidium*, Hook. and Arnott. *L. spinulosa*, Bab. Moore, Nat. Print. Ferns, t. 21. Sowerby, Ferns, t. 12.) Fronds growing upright from the stout, decum-

bent or creeping caudex; stipes nearly as long as the leafy part, stout, brownish-purple below, with a few broad-ovate, pale coloured scales. The fronds vary in height from 2 to 5 feet, they are of a yellowish-green colour, and of a narrow oblonge-lanceolate form, bipinnate. Lower pinnæ distant, nearly opposite, triangular; the upper pinnæ narrow and closer together. Pinnules oblong, pointed at the apex, the lower ones with a short stalk, the upper ones sessile, very much cut or pinnatifid, the lobes deeply serrated, the serratures terminating in short bristle-like teeth. The midvein of the pinnules sends off a vein to each lobe, which is alternately branched, the branches again dividing in a forked manner into the terminal venules. Sori numerous, borne on the short anterior basal venules and forming two rows along the lobes of the pinnules, round, covered by a flat, membranaceous, kidney-shaped indusium; spore-cases brown. Forde bog, found by Mr. C. E. Parker, in 1854, and by myself this year, 1859. P. VI.-IX.

3. **L. dilatata** (*broad prickly-toothed Buckler-Fern.*)—In woods, on banks, and by sides of shaded streams. "Fronds ovate-lanceolate, bipinnate, pinnules pinnate or pinnatifid; segments acutely serrate, spinose-mucronate; indusium with marginal stalked glands; *stipes clothed with long pointed scales, with a dark (nearly black) centre and diaphanous margin*" (*Bab. Man.*). (*Aspidium dilatatum*, E. B. t. 1461. *A. spinulosum*, var. β, Hook. and Arnott. Moore, Nat. Print. Ferns, t. 22. Sowerby's Ferns, t. 13.) Caudex stout and erect, with a thickly scaly crown; stipes about one-third the length of the whole frond, thick below and covered profusely with lanceolate scales, which are light coloured at the edges, but have a well defined dark centre; rachis more sparingly covered with small, pointed, more or less defined, two-coloured scales. Fronds varying in height from 1 or 2 to 6 feet, dark green above, but paler beneath, widely spreading and gracefully arched or drooping, ovate or ovate-lanceolate, bi- or tri-pinnate; the lower pair of pinnæ sometimes spreading out much wider than the rest, and forming, as it were, the base of a triangle, but usually shorter than the pair immediately above them. Pinnæ nearly opposite, the lower ones obliquely triangular, the upper oblong, acute, with the superior and inferior segments nearly equal. Pinnules ovate-oblong, the lower ones stalked, the upper sessile and decurrent, pinnatifid, the divisions terminating in sharp teeth with a bristle-like point. Venation consists of a stout midvein which sends off a strong flexuous vein to each pinnulet, from which a small vein is given off to the marginal lobes; they there divide in a forked manner, and supply a branch to each tooth. The fructification occupies the whole under surface of the frond, the sori being numerous, round, covered by a large

reniform indusium, and placed on the anterior branches of the venules, forming one or two rows along the lobes; spore-cases brown. Side of a small brook, in the valley to the right of the Paignton road, near the railway tunnel. Wood near Milber Down. Woods at Lindridge. Berry Pomeroy woods. Gidleigh, near Chagford, etc. The variety *dumetorum*: "fronds dwarf or dwarfish, oblong-ovate or triangular-ovate, bipinnate; stipes, rachides, and veins beneath clothed with glands; pinnules convex, oblong; scales broad-lanceolate, usually pale, indistinctly two-coloured, fimbriate; sori large, with gland-fringed indusia" (*Moore's Handbook*). Forde bog, near Newton. P. VII.–IX.

4. **Lastrea æmula** (*hay-scented, or triangular pricklytoothed Buckler-Fern.*)—"Fronds triangular or triangular-ovate, spreading, tripinnate, pinnules concave; pinnulets pinnatifid; the mucronately serrate lobes curved upwards; scales of the stipes concolorous, narrow-lanceolate, laciniate, or fimbriate, contorted; indusium margined with minute sessile glands" (*Moore's Handbook*). (*Aspidium spinulosum, var.* γ, Hooker and Arnott. *Lastrea Fœnisecii*, Bab. Moore, Nat. Print. Ferns, t. 27. Sowerby, Ferns, t. 14.) I have never found this species myself in South Devon, but Mr. Moore, in the list of habitats at the end of his 'History of British Ferns' gives "Devil's Tor, Dartmouth," as a station for it. P. VIII. IX.

ATHYRIUM. LADY-FERN.

A. **Filix-fœmina** (*Lady-Fern.*)—In moist shady places, damp woods, and by sides of wooded streams. "Fronds lanceolate, herbaceous, sub-bipinnate or bi-tripinnate; pinnules oblongovate or lanceolate, sessile and distinct, or more or less decurrent and united, toothed, or inciso-pinnatifid, with the lobes toothed, the teeth acute, not spinulose" (*Moore's Handbook*). (Moore's Nat. Print. Ferns, t. 30. Sowerby, Ferns, t. 25. *Aspidium*, E. B. t. 1459: not correct. *Asplenium*, Hooker and Arnott.) Caudex stout and either upright or decumbent, scaly, giving off hard, wiry, dark-coloured roots. Stipes about 1-third or 1-fourth the length of the entire frond, purplish-red or green, covered, especially below, with reddish-brown or very dark-coloured lanceolate or linear scales, the rachis is also sprinkled with smaller and narrower ones. The fronds frequently grow in the vase-like arrangement of *Lastrea Filix-mas*, and vary in height from 1 to 5 feet, they also vary much in their division and form, sometimes broadly and sometimes narrow-lanceolate, sometimes scarcely

FILICES. 159

bipinnate, the pinnules being decurrent, at others almost tripinnate, the segments being divided nearly to the midvein ; the pinnæ are numerous and linear; pinnules linear-oblong or lanceolate, deeply serrate or pinnatifid ; the lobes variously toothed, but never tipped with a bristle. The venation presents many irregularities, but usually consists of a tortuous midvein giving off alternate veins which again divide into venules, the anterior ones bearing a sorus on their upper sides. Fructification occupying the whole back of the frond ; the sori numerous, but very irregular in form, sometimes straight, and in some cases so short as to appear nearly round, but usually of a short, curved, oblong or semilunar outline, covered by a membranaceous iudusium, with a free margin divided into capillary segments. Spore-cases dark brown. Side of a brook running through the valley on the Paignton road. Milber Down. Forde bog. Bradley woods. Berry Pomeroy woods. Side of the stream at Lindridge. Sharpham woods, on the Dart. Holne Chase. Ivybridge. Banks of the Teign at Chagford, etc. The variety *molle* I found in great plenty in Forde bog, about the middle of June this year, 1859, with the fructification considerably advanced. P. VI. VII.

ASPLENIUM. SPLEENWORT.

" Clusters long, straight. Indusium opening towards the central vein or midrib, nearly flat."—*Bab. Manual.*

1. **A. lanceolatum** (*lanceolate Spleenwort.*)—On rocks and walls, not frequent. "*Fronds lanceolate*, bipinnate; pinnules ovate, deeply and sharply toothed or lobed; clusters *short*, nearly marginal" (*Bab. Man.* p. 425). (E. B. t. 240. Moore, Nat. Print. Ferns, t. 35 B. Sowerby, Ferns, t. 27 : very incorrect. Moore, Handbook, p. 167.) Caudex with many stout branching roots, short, either upright or decumbent, covered with shining, brown, awl-shaped scales. Stipes one-third, sometimes one-half, as long as the whole frond, of a rich chestnut colour below, gradually shading off into the greener rachis, slightly scaly ; the upper side of the rachis has a slightly raised margin, more or less distinct in different specimens, and is sparsely supplied with small slender hairs. Fronds from 2 or 3 to 12 or more inches high, stiff, smooth, upright or drooping, lanceolate, bipinnate, of a fresh, bright green colour. The lowest pair of pinnæ usually much shorter than those immediately above, but sometimes they spread out much wider than any of the rest, and impart to the frond a deltoid or triangular, instead of a lanceolate outline. Pinnæ either nearly opposite or alternate, narrowing from the

base to the point; segments of the pinnæ varying much in the extent of their division and form, sometimes obovate, or inclining to a square outline, and frequently in vigorous fronds pinnatifid, with obovate sharply-toothed lobes; the pinnules bear a flexuous midvein which gives off alternate veins, the lowest anterior one being distributed to the principal lobe, giving off venules to each tooth; the other veins, either branched or undivided, proceed to the marginal teeth. Fructification generally sprinkling the whole dorsal surface; the sori are oblong, and are situated on the anterior side of the venules, occupying rather the middle of the lobes than the centre of the pinnules; they are covered by a white, oblong indusium, which has a free, wavy margin; the sori are at first distinct from each other, but in process of ripening they become confluent, and form irregular patches on the lobes. Spore-cases bright brown. Crevices of rocks facing the sea, near Hope's Nose. Rocky bank in a steep lane leading from Barton into the Newton road. P. v.–viii.

2. **A. Adiantum-nigrum** (*black Maidenhair Spleenwort.*)— On hedgebanks, in crevices of rocks, and on old walls, very common. "*Fronds ovate-triangular*, twice or thrice pinnate, about *as long as the stipes*, pinnæ and pinnules triangular, *ultimate subdivisions blunt*, sharply toothed, *clusters long*, central" (*Bab. Man.* p. 426). (Sm. E. B. t. 1950. Moore, Nat. Print. Ferns, t. 36. Moore, Handbook, p. 170. Sowerby, Ferns, t. 28.) Caudex stout, short and tufted, bearing lanceolate, pointed scales, and sending down numerous branching roots. The stipes, which is of a shining dark purplish-brown colour, is usually as long as, but sometimes longer than, the leafy portion of the frond. In some specimens I picked this year, growing on hedgebanks among long grass, the stipes were nearly double the length of the leafy portion; the dark colour of the stipes extends to the back of the rachis. Length of the whole fronds from 3 or 4 to 18 or 20 inches, they present a rigid, leathery appearance, and are of a dark shining green on the upper surface, but paler beneath; their form is deltoid or ovate-triangular, always tapering to a sharp-pointed apex; they are twice, thrice, and sometimes almost four times pinnate. Pinnæ obliquely triangular, thinning off at the apex, the lowest pair nearly opposite, and generally, though not always, longer than the rest; the pinnæ usually point upwards towards the apex of the frond; they differ very much in their division; the ultimate lobes are unequally toothed, the teeth being more or less narrowed or obtuse, but always ending in a point. The venation varies with the division of the frond; the fructification, however, is always situated near the giving off of the branches from the midvein, and so occupies the centre of the pinnules. Sori linear and crowded, covered with a whitish indu-

sium, becoming confluent as they arrive at maturity, and occupying the whole under surface; spore-cases round, shining brown. This Fern differs much in appearance according to the situation in which it grows, and has accordingly been divided by many botanists into several varieties. Very abundantly distributed on walls, rocks, and hedge-banks. A very small variety is found on old walls, with the pinnæ much crowded together, and their segments very slightly divided. P. v.–viii.

3. **A. marinum** (*sea Spleenwort.*)—In crevices of rock and rocky caves near the sea. "Fronds linear, simply pinnate; pinnæ stalked, ovate or oblong, serrate, unequal, and wedge-shaped at the base" (*Bab. Man.* p. 426). (Sm. E. B. t. 392. Moore, Nat. Print. Ferns, t. 38. Sowerby, Ferns, t. 29. Moore, Handbook, p. 177.) Caudex short, fixed firmly by long and numerous slender wiry roots; stipes shorter than the leafy part of the frond, purplish-brown, smooth; the rachis is also slightly coloured below, but green and winged above. Fronds from 2 or 3 to 6 or 12 inches long, or even more, broadly linear, but tapering to the summit. Pinnæ stalked, oblong or ovate, oblique from the lengthening of basal angle anteriorly, the margins crenated and serrate. Venation consisting of a strong midvein, which gives off lateral veins which are again forkedly divided. Sori situated on the anterior side of the venules, linear, covered by a permanent undivided indusium, they range on each side of the midvein, and form two rows of oblique lines at the back of each of the pinnæ. Spore-cases numerous, brown. Rocks at Hope's Nose (plants very small). Caves in the rocks around Torbay. Formerly plentiful on the rocks at Paignton, but the habitat has been destroyed. P. vi.–ix.

4. **A. Trichomanes** (*common Maidenhair Spleenwort.*)—On rocks and old walls, plentifully distributed. "Fronds linear pinnate; pinnæ roundish-ovate, crenate; *veins forked below the clusters*" (*Bab. Man.* p. 426). (Sm. E. B. t. 576. Moore, Nat. Print. Ferns, t. 39. Sowerby, Ferns, t. 30. Moore, Handbook, p. 181.) Caudex very short; roots wiry, long, and branching. Stipes exceedingly short, smooth, of a rich chestnut colour, or very dark brown, sometimes nearly black, rounded posteriorly, but flat in front, and having a raised line on each side; rachis dark coloured like the stipes; fronds varying from 2 to 14 or 15 inches in length, linear and simply pinnate; pinnæ numerous, roundish-ovate or oblong, blunt at the apex, deep full green and glossy, attached to the rachis by a very short stalk formed by the narrowing of the wedge-shaped base, their margins are either crenated or entire. The venation comprises a midvein which gives off forked veins, upon the anterior venules of which the sori are borne. Sori numerous, linear, covered by an indusium,

and situated above the point where the veins fork, becoming confluent when they attain to maturity; the fructification is distributed over the whole back of the frond. Rocks near the sea by Daddyhole Plain. Babbicombe Down. Old walls about Torquay and Cockington. Rocks near Hope's Nose. Wall of Kingskerswell churchyard, in great abundance. Walls at Paignton, etc. P. v.-ix.

5. **A. Ruta-muraria** (*rue-leaved Spleenwort, or Wall-Rue.*) —In crevices of rocks and on old walls, frequent. " Fronds bipinnate; *pinnules rhomboid* wedge-shaped, notched or toothed at at the end; *indusium jagged*" (*Bab. Man.* p. 426). (Sm. E. B. t. 150. Moore, Nat. Print. Ferns, t. 41 A. Sowerby, Ferns, t. 32. Moore, Handbook, p. 188.) Caudex short and tufted, with numerous branching fibres; fronds growing in tufts, from 1 to 6 or 8 inches long, borne on a smooth purplish-brown stipes fully as long as or longer than themselves; the fronds are of a deep green colour, and of a somewhat leathery substance, deltoid in outline, and usually bipinnate: in young and starved plants the fronds are occasionally simply pinnate, with roundish or kidney-shaped pinnæ. Pinnæ and pinnules alternate; pinnules rhomboidal, wedge-shaped at the base, the upper margin irregularly toothed or serrate. Venation consisting of veins running from the base of the pinnules, and branching above in a forked manner, sending a venule to every tooth and serrature, there being no apparent midvein. The sori are situated on the inner side of the veins, occupying the centre of the pinnæ or pinnules, they are linear in form, and covered by a narrow white indusium with a wavy or jagged margin which is however not distinguishable after the earlier stages of fructification, the sori in the latter stages becoming confluent. The spore-cases are dark brown. Rocks near the chasm at Daddyhole Plain. Rocks on Babbicombe Down. Walls about Torquay, Cockington, and Shiphay. Old walls, near Paignton. Kingskerswell churchyard wall, etc. P. v.-ix.

SCOLOPENDRIUM. HART'S-TONGUE FERN.

" Clusters long, straight, two together. Indusia of each pair opening towards each other." *Bab. Man.* p. 418.

S. vulgare (*common Hart's-tongue Fern.*)—On old walls and ruins, on shady hedge-banks, in woods and thickets, on the sides of wells, and in moist shady places generally. " Fronds oblong-strap-shaped, smooth, simple, with a cordate base, stipes shaggy" (*Bab. Man.* p. 427). (Sm. E. B. t. 1150. Moore, Nat. Print. Ferns, t. 42. f. 1. Sowerby, Ferns, t. 35. Moore, Handbook,

p. 197.) Fronds from 5 or 6 inches to 2 feet long, borne on a shortish, purplish-brown, scaly stipes, from a short tufted caudex: they are long-lanceolate, broadly linear, or oblong strap-shaped, entire or slightly wavy at the margin, and cordate at the base; the veins appear to spring from the midrib, forking two or three times near the base, and running parallel to each other, send venules to near the margin, where they terminate in little club-shaped apices. Sori long and straight, set obliquely, growing two together, at length running into each other, each one covered by a narrow indusium, the two indusia opening towards each other by parting down the middle of the twin sori. Spore-cases reddish-brown, numerous; the fructification most abundant at the upper part of the frond. The fronds of this Fern vary very much in appearance, and frequently put on most fantastic shapes: many of these abnormal forms are described as varieties, and Mr. Moore, in his Handbook, gives some 60 of them under distinct Latin names. I have this year found entire normally-shaped, crisped, bifid and multifid fronds growing from one and the same caudex. Very abundantly distributed throughout the whole district. P. VII. VIII.

CETERACH. SCALE-FERN.

"Lateral veins anastomosing, clusters attached to their middle on the side next the midrib, except in the lowest; indusium (?) a narrow nearly erect membrane on the back of the vein. Whole back of the frond covered with chaffy scales."—*Bab. Man.* p. 418.

C. officinarum (*common Scale-Fern.*)—On old walls and rocks. "Fronds coriaceous, narrow-lanceolate, sinuato-pinnatifid, often pinnate below; segments oblong, obtuse, entire or sinuately lobed, densely scaly beneath" (*Moore, Handbook,* p. 213). (Moore, Nat. Print. Ferns, t. 43 A. Sowerby, Ferns, t. 36. *Scolopendrium Ceterach,* Sm. E. B. t. 1244. *Asplenium Ceterach,* Linn. Moore's Handbook, p. 214.) Caudex short, covered with dark brown scales, and giving off fibrous roots; stipes dark coloured below, very short, covered with broadly lanceolate scales. Fronds from 1 to 8 or more inches long, dark green above, covered entirely beneath with dense tawny scales, linear-lanceolate and deeply pinnatifid; the lobes are bluntly oblong, and are not divided down to the rachis, but unite with each other at the base. The venation is only to be distinguished at a very early stage of the fructification: it consists of a wavy midvein giving off lateral veins, which are irregularly forked, and send anastomosing venules which terminate near the margin. Sori oblong or linear, situated on the upper side of the anterior venules. Indusium said to be obsolete, but traceable in an early stage of the development of the frond. The whole fructification is very much obscured by the dense and numerous scales covering the under surface. Formerly very plentiful about the rocks of Torquay, but it has of late years been mercilessly rooted up to satisfy the cravings of fern-cultivators. Rocks about Babbicombe Down. Churchyard walls at Chudleigh and Ilsington, *Fl. D.* P. IV.-X.

BLECHNUM. HARD-FERN.

" Capsules in a continuous line parallel to the midrib, upon a longitudinal anastomosing part of the transverse veins, covered by a continuous scarious indusium."—*Bab. Man.*, p. 418.

Blechnum Spicant (*common Hard-Fern.*)—In damp, stony, and heathy places. "Barren fronds pectinate-pinnatifid with broadly linear rather obtuse pinnæ; fertile frond pinnate, with linear acute pinnæ" (*Bab. Man.*, p. 427). (Moore, Nat. Print. Ferns, t. 43 C. *B. boreale*, Sm. E. B. t. 1159; Sowerby, Ferns, t. 37; Hook. and Arn. Br. Fl.; Bab. Man. *Lomaria Spicant*, Newman.) Caudex stout, either upright or decumbent, having stoutish branched roots, and being covered with tawny-brown narrow-lanceolate scales. Stipes of barren fronds short, those of the fertile fronds much longer, dark purplish-brown. Barren fronds from 6 to 18 inches long, dark-green, usually spreading, pinnatifid, with the linear-oblong lobes arranged on each side the rachis like the teeth of a comb, the lobes at the apex of the frond confluent. Fertile fronds taller, often exceeding 2 feet in height, pinnate below, and with much narrower lobes. Venation of the barren fronds a stout midvein, which sends off veins which are once or twice forked, the venules ending near the margin in a club-shaped head; in the fertile fronds the lateral veins are alternate, and proceed upward about halfway to the margin, then they make an abrupt turn and run parallel with the midvein, each anastomosing with the one above it, and forming apparently a longitudinal vein on each side the midvein. Sori linear, one on each side of the midrib, extending the entire length of the pinnæ, covered by a continued membranous indusium, sori becoming ultimately confluent. Forde bog. Bovey Heath. Ivybridge. Moor at Chagford. P. VI. VII.

PTERIS. BRAKES OR BRACKEN.

" Sori indusiate, marginal, linear, continuous or interrupted;

FILICES. 165

the *receptacles* linear transverse, uniting the apices of the veins. Indusium of the same form, membranaceous. Veins simple or forked from a central costa; venules free."—*Moore, Handbook*, p. 223.

P. aquilina (*common Brakes or Bracken.*)—On heaths and downs, in woods and shady places, frequent. " Fronds tripartite, branches bipinnate, pinnules linear-lanceolate, the lower ones usually pinnatifid; segments oblong-obtuse" (*Bab. Man.*, p. 427). (E. B. t. 1679. Moore, Nat. Print. Ferns, t. 44. Sowerby, Ferns, t. 38.) "Fronds annual, 1 to 5 feet high, very much divided, with spreading branches. Capsules attached to the marginal vein, lying upon a fine membrane, and covered by the membranous continuation of the epidermis. Inferior pinnules pinnatifid, or sinuate, or entire" (*Bab. Man.*, p. 427). Very abundantly distributed everywhere. P. VII.

ADIANTUM. MAIDENHAIR-FERN.

"Capsules marginal, oblong or roundish, covered by distinct reflexed portions of the margin of the frond."—*Bab. Man.*, p. 418.

A. Capillus-Veneris (*common Maidenhair.*)—In crevices of damp rocks, and in moist caves, more especially near the sea. "Frond irregular; branches and roundish wedge-shaped, lobed, thin pinnules alternate; lobes of the fertile pinnules terminated by a transversely linear-oblong reflexed lobe covering several roundish clusters: sterile lobes serrate." (*Bab. Man.*, p. 428). (E. B. t. 1564. Moore, Nat. Print. Ferns, t. 45; Sowerby, Ferns, t. 40; Moore, Handbook, p. 230.) "Fronds bi-tripinnate; pinnules alternate, glabrous, membranaceous, obliquely and broadly wedge-shaped, or roundish, with a truncate base, attached by capillary stalks, the superior margin lobed, the sterile lobes dentate, the fertile, obtuse or truncate; sori tranversely oblong, often occupying the whole width of the lobes; stipes and rachis ebony-black, smooth, glossy" (*Moore, Handbook*, p. 229). For a detailed description of this most delicate and lovely Fern, see Moore's Handbook. In the crevices of some wet rocks at Mudstone Bay, near Brixham. Caves at Brixham. P. V.-IX.

HYMENOPHYLLUM. FILMY-FERN.

"Capsules on a narrow subclavate receptacle, within a two-valved involucre of the same texture with the frond."—*Bab. Man.* p. 418.

1. **H. tunbridgense** (*Tunbridge Filmy-Fern.*)—On damp rocks, and amongst moss, in moist and shady places. "Fronds pellucido-membranaceous, ovate or oblong, more or less elongated, pinnate; pinnæ subvertical, pinnatifid, decurrent, forming a wing to the rachis; segments linear, undivided or bifid, and as well as the upper margin of the roundish valves of the axillary solitary sessile compressed involucres, spinulosely serrate" (*Moore, Handbook*, p. 261). (E. B. t. 162. Moore, Nat. Print. Ferns, t. 49 A. Sowerby, Ferns, t. 42. Moore, Handbook, p. 262.) This Fern and the following species have very much the appearance and habit of some of the Mosses, for which they may easily be mistaken, without a careful examination. Bickleigh Vale. Beckey Fall. P. VII.

2. **H. unilaterale** (*Wilson's Filmy-Fern.*)—In similar localities to the last. "Fronds pinnate, pinnæ recurved; segments linear, undivided or bifid, spinosely serrate; *involucre inflated, entire*; rachis slightly bordered" (*Bab. Man.*, p. 428). (Moore, Handbook, p. 264. Moore, Nat. Print. Ferns, t. 49 B. Sowerby, Ferns, t. 43. *H. Wilsoni*, E. B. S. t. 2686; Hook. and Arn. Brit. Fl.; Bab. Man.) Very much resembling the preceding, but distinguished chiefly by the involucre being stalked instead of sessile, by its being longer and ovate instead of rounded, and by its edges never being serrate. Tors on Dartmoor. Beckey Fall, *Mr. C. E. Parker.* P. VII.

OSMUNDA. ROYAL FERN.

"Capsules clustered, arranged in a branched spike terminating the frond."—*Bab. Man.* p. 419.

O. regalis (*Royal or Flowering Fern, or Osmund Royal.*)— In wet and boggy places, and in the damp borders of woods. Fronds bipinnate, pinnæ opposite; the pinnules are oblong and nearly entire, auricled to a greater or lesser degree at the base; fructification crowning the fertile fronds in terminal bipinnate panicles. (E. B. t. 209. Moore, Nat. Print. Ferns, t. 50. Sowerby, Ferns, t. 44. Moore, Handbook, p. 269.) The grandest and most stately in its growth of all our British Ferns, the fronds varying in height from 2 or 3 to 8 feet, but sometimes rising considerably higher. I have gathered fronds in Holne Chase which measured 10, 12, and 13 feet. The fructification is confined to the upper pinnæ, which are converted into a clustered panicle, the spore-cases covering the pinnules either wholly or in part. For a detailed description of the specific characters, see *Moore's Handbook.* Forde bog. Bovey Heath. Holne Chase, very plentiful and of most luxuriant growth. Holystreet and Gidleigh, near Chagford. Ivybridge, etc. P. vii.-ix.

Ord. II. OPHIOGLOSSACEÆ.

BOTRYCHIUM. MOONWORT.

"Capsules distinct, disposed in a compound spike, attached to a pinnate or bipinnate frond."—*Bab. Man.* p. 419.

B. Lunaria (*common Moonwort.*)—In dry, open, and elevated pastures. "Fronds solitary; barren branch oblong, pinnate; pinnæ lunate or fan-shaped, the margin jagged or crenate" (*Moore*, p. 271). (Moore, Nat. Print. Ferns, t. 51. Moore, Handbook, p. 272. . Sowerby, Ferns, t. 45. *Osmunda Lunaria*, Linn. E. B. t. 318.) "Height from 3 to 6 inches. Pinnæ with veins radiating from the petiole, sometimes deeply notched. Fronds usually solitary, but sometimes two on the same stalk" (*Bab.* p. 429). Haldon, *Miss A. Griffith*. Mr. Moore, in his 'Popular History of British Ferns,' gives "by the Dart" as a habitat, but does not specify where. I have never found this Fern in Devonshire myself. P. vi. vii.

OPHIOGLOSSUM. ADDER'S-TONGUE.

"Capsules connate, disposed in a simple distichous spike, attached to an undivided frond."—*Bab.* p. 419.

O. vulgatum (*common Adder's-tongue.*)—In moist pastures and meadows. Frond ovate-obtuse, with a club-shaped spike rising higher than the leafy expansion. (E. B. t. 108. Moore, Nat. Print. Ferns, t. 51 B. Moore, Handbook, p. 276. Sowerby, Ferns, t. 46.) Fronds from 2 or 3 to sometimes 12 inches high, on a smooth cylindrical and hollow stipes; the barren branch leaf-like, smooth, entire, sessile, broad-ovate, either obtuse or acute; the fertile branch or spike erect, on a longer or shorter stalk; the fructifications are arranged on the margins of the spike, the thecæ or spore-cases are round and smooth, sessile, arranged in two rows, one on each margin, bursting transversely with two valves. In a meadow near Torre Abbey, but the habitat has been destroyed this year by building. At the base of the mound at Ellacombe, *Mr. Earle*. Meadows about Exeter. P. v. vi.

EQUISETACEÆ.

"Leafless branched plants, with a striated fistular stem; joints sheathed above each joining. Sporules surrounded by elastic clavate filaments, and enclosed in capsules arising from the peltate scales of terminal cones or spikes. Venation straight. Cuticle abounding in silex. Only one genus."—*Bab. Man.* p. 414.

EQUISETUM. HORSE-TAIL.

1. **E. fluviatile** (*great Water Horse-tail.*)—In wet places and by sides of rivers and pools, frequent. Sterile stems almost smooth, with about 30 striæ and branches, which are simple and nearly erect, from 3 to 6 feet high. Fertile stem unbranched, stout, about a foot high, appearing before the sterile ones, having numerous pale brown sheaths, and bearing a large spike. (E. B. t. 2022.) *E. Telmateia*, Bab. Petit Tor. Watcombe. Maidencombe. Shaldon. Cliffs at Dawlish. (Ilsington, *Fl. D.*) P. IV.

2. **E. arvense** (*corn Horse-tail.*)—In damp meadows and by roadsides, abundant. "Frond attenuated upwards, sterile stem slightly scabrous, with 12 to 14 furrows, teeth of the sheath lanceolate-subulate 1-ribbed to the point, branches simple erecto-patent, fertile stem without branches, its sheaths remote loose" (*Brit. Flor.* p. 598). The sterile and fertile stems are always distinct, the latter appearing first. (E. B. t. 2020.) Ansti's Cove lane. Teignmouth road, etc. P. IV.

3. **E. limosum** (*smooth naked Horse-tail.*)—In watery places and ditches. Sterile and fertile stems similar; stems nearly smooth, striate; striæ about 16 or 18; branches nearly erect, simple, whorled. Stem from 2 to 4 feet high, spike blunt. (E. B. t. 929.) Petit Tor, near St. Marychurch. P. VI. VII.

4. **E. palustre** (*marsh Horse-tail.*)—In wet and boggy places, frequent. Sterile and fertile stems similar. "Stems with 4 to 8 deep furrows, branched throughout; sheaths loose, pale, with acute wedge-shaped teeth, tipped with brown, and membranous at the edges. Barren stem whip-shaped at the end. Spike blunt. Sheaths coloured like the stem, or paler" (*Bab. Man.* p. 416). Watcombe. Common in boggy places, *Fl. D.* P. VI. VII.

LYCOPODIACEÆ.

" Leafy plants with simple imbricated leaves, or stemless, with erect subulate leaves. Fructification of axillary sessile capsules, with two or three valves, and no ring, including minute powdery matter or sporules."—*Bab. Man.* p. 430.

LYCOPODIUM. CLUB-MOSS.

1. **L. clavatum** (*common Club-moss.*)—In moors and heathy places, common. Stem long and prostrate, with short ascending branches ; leaves scattered and incurved. Spikes pale yellow, on long stalks. Haldon. Many parts of Dartmoor. (E. B. t. 224.) P. VII. VIII.

2. **L. Selago** (*Fir Club-moss.*)—In mountainous heaths. "*Leaves in eight rows, crowded*, uniform, linear-lanceolate, acuminate, capsule not spiked, but in the axils of the common leaves ; stem erect, forked, level-topped" (*Bab. Man.* p. 431). Stems varying from 3 or 4 to 6 or 8 inches in height, and branching in two-forked manner, usually upright, bearing commonly at their extremities a few deciduous buds, which fall to the ground on separation, and there vegetate. Leaves thick, of a shining dark-green. Frequent in Dartmoor. Woodbury Hill. (E. B. t. 233.) P. VI.-VIII.

3. **L. inundatum** (*marsh Club-moss.*)—In boggy heaths. "Leaves scattered, linear, acute, turned upwards ; spikes terminal, sessile, leafy, solitary, upon short erect branches. Stem short, prostrate, rooting. Branches few, simple, short, erect, fertile" (*Bab. Man.* p. 431). More diminutive than the two former, its prostrate simple stems being about 2 or 3 inches long, and growing close to the surface of the ground. Spore-cases in the axils of the bracts. Forde bog. Bovey Heath. (E. B. t. 369.) P. VIII. IX.

THE FOLLOWING SPECIES HAVE BEEN ACCIDENTALLY OMITTED.

Ord. XXXIV. PARONYCHIACEÆ, page 46.

Corrigiola littoralis (*Sand Strapwort.*)—South-western coast of England. Leaves alternate, fleshy, glaucous; flowers white, calyx 5-partite, petals five, same length as the calyx. Slapton sands, and near Start Point, *Fl. D.* (E. B. t. 668.) A. VII. VIII.

Ord. CVII. GRAMINEÆ, page 146.

Sesleria cœrulea (*blue Moor-Grass.*)—Various places in Dartmoor. (E. B. t. 1613.) P. IV.–VI.

ERRATUM.

Page 14. In the description of DIPLOTAXIS TENUIFOLIA instead of "Leaves a yellowish green" *read* " Leaves glaucous."

TABLE

OF THE

NATURAL ORDERS OF FLOWERING PLANTS,

SHOWING THE PAGES WHERE THE SPECIES UNDER EACH ORDER ARE TO BE FOUND.

This mark † placed before the name of an Order denotes that there are no Species belonging to that Order in our Flora.

Order	Page	Order	Page
1. Ranunculaceæ	1–4	25. Rhamnaceæ	29
2. Berberidaceæ	5	26. Leguminosæ	29–38
3. Nymphæaceæ	5	27. Rosaceæ	38–44
4. Papaveraceæ	5, 6	28. Onagraceæ	44
5. Fumariaceæ	7	29. Haloragaceæ	45
6. Cruciferæ	7–15	30. Lythraceæ	45
7. Resedaceæ	15	31. Tamaricaceæ	45
8. Cistaceæ	15, 16	†32. Cucurbitaceæ	46
9. Violaceæ	16, 17	33. Portulaceæ	46
10. Droseraceæ	17	34. Paronychiaceæ	46, 47
11. Polygalaceæ	17	35. Crassulaceæ	47, 48
†12. Frankeniaceæ	18	36. Grossulariaceæ	48
†13. Elatinaceæ	18	37. Saxifragaceæ	48
14. Caryophyllaceæ	18–22	38. Umbelliferæ	49–56
15. Linaceæ	22, 23	39. Araliaceæ	56
16. Malvaceæ	23	40. Cornaceæ	57
17. Tiliaceæ	24	41. Loranthaceæ	57
18. Hypericaceæ	24, 25	42. Caprifoliaceæ	58
19. Aceraceæ	26	43. Rubiaceæ	59, 60
20. Geraniaceæ	26–28	44. Valerianaceæ	61
†21. Balsaminaceæ	28	45. Dipsaceæ	62
22. Oxalidaceæ	28	46. Compositæ	63–77
†23. Staphylaceæ	28	47. Campanulaceæ	77, 78
24. Celastraceæ	28	†48. Lobeliaceæ	78

ORDERS OF FLOWERING PLANTS.

Order	Page
49. Vacciniaceæ	78
50. Ericaceæ	79
†51. Pyrolaceæ	80
†52. Monotropaceæ	80
53. Aquifoliaceæ	80
54. Oleaceæ	80
55. Apocynaceæ	81
56. Gentianaceæ	81
†57. Polemoniaceæ	82
58. Convolvulaceæ	82
59. Boraginaceæ	83–86
60. Solanaceæ	86
61. Orobanchaceæ	87
62. Scrophulariaceæ	88–94
63. Labiatæ	94–102
64. Verbenaceæ	102
65. Lentibulariaceæ	102
66. Primulaceæ	103–105
67. Plumbaginaceæ	105
68. Plantaginaceæ	106, 107
†69. Amaranthaceæ	107
70. Chenopodiaceæ	107–110
71. Scleranthaceæ	110
72. Polygonaceæ	110–113
73. Thymelaceæ	113
74. Santalaceæ	114
†75. Aristolochiaceæ	114
†76. Empetraceæ	114
77. Euphorbiaceæ	114, 115
78. Callitrichaceæ	115
79. Ceratophyllaceæ	116
80. Urticaceæ	116, 117
81. Ulmaceæ	117
†82. Elæagnaceæ	118
83. Myricaceæ	118
84. Betulaceæ	118
85. Salicaceæ	119–122
86. Cupuliferæ	122, 123
87. Coniferæ	123
88. Hydrocharidaceæ	124
89. Orchidaceæ	124–127
90. Iridaceæ	127
91. Amaryllidaceæ	128
92. Dioscoreaceæ	128
†93. Trilliaceæ	129
94. Liliaceæ	129, 130
95. Melanthaceæ	130
†96. Restiaceæ	131
97. Juncaceæ	131–133
98. Butomaceæ	133
99. Alismaceæ	134
100. Juncaginaceæ	134
101. Typhaceæ	135
102. Araceæ	135
103. Orontiaceæ	136
104. Pistiaceæ	136
105. Naiadaceæ	137, 138
106. Cyperaceæ	139–143
107. Gramineæ	143–152

LATIN INDEX TO THE GENERA.

The names printed in Italics are Synonyms.

	Page		Page
Acer	26	Anthriscus	55
Achillea	77	Anthyllis	30
Acorus	136	Antirrhinum	92
Adiantum	165	Apargia	64
Adoxa	56	Apium	50
Ægopodium	51	Aquilegia	5
Æthusa	53	Arabis	8
Agraphis	129	Arctium	67
Agrimonia	42	Arenaria	20
Agrostemma	19	Armeria	105
Agrostis	145	Armoracia	9
Aira	145	Arrhenatherum	146
Ajuga	97	Artemisia	71
Alchemilla	41	Arum	135
Alisma	134	*Arundo*	150
Alliaria	12	Asparagus	129
Allium	130	Asperula	60
Alnus	118	*Aspidium*	156
Alopecurus	143	Asplenium	159
Ammophila	145	Aster	73
Anagallis	104	Athyrium	158
Anchusa	85	Atriplex	108
Anemone	2	Avena	150
Angelica	54		
Anthemis	77	Ballota	97
Anthoxanthum	143	Barbarea	8

174 LATIN INDEX TO THE GENERA.

	Page		Page
Bartsia	89	Ceratophyllum	116
Bellis	75	Ceterach	163
Berberis	5	Chærophyllum	55
Beta	107	Cheiranthus	7
Betonica	99	Chelidonium	6
Betula	118	Chenopodium	107
Bidens	70	Chlora	82
Blechnum	164	Chrysanthemum	76
Borago	86	*Chrysocoma*	72
Botrychium	167	Chrysosplenium	49
Brachypodium	151	Cichorium	67
Brassica	13	Circæa	44
Briza	148	*Cistus*	16
Bromus	149	Clematis	1
Bunium	51	*Clinopodium*	101
Bupleurum	52	Cnicus	68
Butomus	133	Cochlearia	10
		Colchicum	130
Cakile	11	Conium	54
Calamagrostis	145	Convolvulus	82
Calamintha	100	*Conyza*	75
Callitriche	115	Cornus	57
Calluna	79	*Coronopus*	13
Caltha	4	Corrigiola	170
Calystegia	82	Corydalis	7
Campanula	77	Corylus	123
Capsella	12	Cotyledon	47
Cardamine	8	Crambe	15
Carduus	68	Cratægus	43
Carex	141, 142	Crepis	65
Carlina	69	Crithmum	54
Carpinus	123	Cuscuta	83
Castanea	122	Cynoglossum	86
Caucalis	56	Cynosurus	148
Centaurea	70		
Centranthus	61	Dactylis	148
Centunculus	105	Daphne	113
Cerastium	21	Daucus	56

Page
Dianthus 18
Digitalis 91
Diplotaxis 14
Dipsacus 62
Draba 10
Drosera 17

Echium 83
Eleocharis 139
Elymus 151
Endymion 130
Epilobium 44
Epipactis 124
Equisetum 168
Erica 79
Eriophorum . . . 140
Erodium 27
Ervum 37
Eryngium 49
Erysimum 12
Erythræa 81
Euonymus 28
Eupatorium . . . 72
Euphorbia 114
Euphrasia 90

Fagus 122
Fedia 61
Festuca 148
Ficaria   2
Filago 72
Fœniculum 53
Fragaria 40
Fraxinus 80
Fumaria   7

Galanthus 128
Galeobdolon . . . 98

         Page
Galeopsis 98
Galium 59
Gastridium . . . 144
Genista 30
Gentiana 81
Geranium 26
Geum 39
Glaucium   6
Glaux 103
Glechoma 100
Glyceria 147
Gnaphalium . . . 72

Habenaria 126
Hedera 57
Hedypnois 64
Hedysarum 35
Helianthemum . . 15
Helleborus   4
Helminthia 63
Helosciadum . . . 50
Heracleum 54
Hieracium 66
Hippocrepis . . . 35
Hippuris 45
Holcus 146
Hordeum 151
Humulus 117
Hyacinthus . . . 130
Hydrocharis . . . 124
Hydrocotyle . . . 49
Hymenophyllum . 165
Hyoscyamus . . . 86
Hypericum 24
Hypochœris . . . 64

Ilex 80
Illecebrum 46

LATIN INDEX TO THE GENERA.

	Page		Page
Inula	75	Lysimachia	104
Iris	127	Lythrum	45
Isolepis	139		
		Malachium	22
Jasione	78	Malva	23
Juncus	131, 132	Marrubium	100
		Matricaria	76
Knautia	62	Medicago	31
Kobresia	140	Melampyrum	90
Kœleria	146	Melica	146
Koniga	10	Melilotus	31
		Melittis	101
Lamium	98	Mentha	95
Lapsana	66	Menyanthes	82
Lastrea	156	Mercurialis	114
Lathræa	87	Milium	144
Lathyrus	37	Mœnchia	21
Lavatera	23	Molinia	146
Lemna	136	Montia	46
Leontodon	65	Myosotis	84
Leonurus	97	Myrica	118
Lepidium	13	Myriophyllum	45
Lepturus	152		
Ligustrum	80	Narcissus	128
Linaria	92	Nardus	143
Linosyris	72	Narthecium	133
Linum	22	Nasturtium	9
Listera	124	Neottia	125
Littorella	106	Nepeta	100
Lithospermum	83	Nuphar	5
Lolium	152	Nymphæa	5
Lonicera	58		
Lotus	34	Œnanthe	52
Luzula	132, 133	Onobrychis	35
Lychnis	19	Ononis	30
Lycopodium	169	Onopordum	69
Lycopsis	85	Ophioglossum	167
Lycopus	94	Ophrys	127

LATIN INDEX TO THE GENERA.

	Page		Page
Orchis	125	Pyrethrum	76
Origanum	96	Pyrus	43
Ornithopus	35		
Orobanche	87	Quercus	122
Orobus	38		
Osmunda	166	Radiola	23
Oxalis	28	Ranunculus	2
		Raphanus	15
Papaver	5	Reseda	15
Parietaria	116	Rhamnus	29
Pedicularis	91	Rhinanthus	90
Peplis	45	Rhyncospora	139
Petasites	73	Ribes	48
Petroselinum	50	Rosa	42
Phalaris	144	*Rottboellia*	152
Phleum	144	Rubia	59
Phragmites	150	Rubus	40
Picris	63	Rumex	112
Pimpinella	51	Ruppia	138
Pinguicula	102	Ruscus	129
Pinus	123		
Plantago	106	Sagina	19
Poa	147	Sagittaria	134
Polycarpon	46	Salicornia	109
Polygala	17	Salix	119, 120, 121
Polygonum	110, 111	Salsola	110
Polypodium	153	Salvia	95
Polystichum	154	Sambucus	58
Populus	121	Samolus	105
Potamogeton	137	Sanguisorba	41
Potentilla	41	Sanicula	49
Poterium	41	Saponaria	18
Primula	103	Sarothamnus	30
Prunella	101	Saxifraga	48
Prunus	38	Scabiosa	62
Psamma	145	Scandix	55
Pteris	164	Schœnus	139
Pulicaria	75	Scilla	130

N

	Page		Page
Scirpus	140	Symphytum	85
Scleranthus	110		
Sclerochloa	147	Tamarix	45
Scolopendrium	162	Tamus	128
Scrophularia	91	Tanacetum	71
Scutellaria	101	Taxus	123
Sedum	47	Teesdalia	11
Sempervivum	47	Teucrium	97
Senebiera	13	Thalictrum	1
Senecio	74	Thesium	114
Serapias	124	Thlaspi	10
Serratula	67	Thrincia	64
Serrafalcus	149	Thymus	96
Sesleria	170	Tilia	24
Sherardia	60	Torilis	56
Sibthorpia	93	*Tormentilla*	41
Silaus	53	Tragopogon	63
Silene	18	Trichonema	127
Silybum	68	Trifolium	32
Sinapis	14	Triglochin	134
Sison	51	Trigonella	32
Sisymbrium	11	Trinia	50
Sium	53	Triodia	148
Smyrnium	55	*Trisetum*	150
Solanum	86	Triticum	151
Solidago	74	Tussilago	73
Sonchus	64	Typha	135
Sparganium	135		
Spartium	30	Ulex	29
Spergula	47	Ulmus	117
Spergularia	46	Urtica	116
Spiræa	39	Utricularia	102
Spiranthes	125		
Stachys	99	Vaccinium	78
Statice	105	Valeriana	61
Statice Armeria	105	Verbascum	93
Stellaria	20	Verbena	102
Suæda	109	Veronica	88

	Page		Page
Viburnum	58	Viscum	57
Vicia	36		
Vinca	81	Zanichellia	138
Viola	16	Zostera	138

INDEX

TO THE

POPULAR ENGLISH NAMES.

	Page		Page
Abele	121	Beaked Parsley	55
Adder's-tongue	167	Bedstraw	59
Agrimony	42	Beech	122
Alder	118	Bee Orchis	127
Alder Buckthorn	29	Beet	107
Alexanders	55	Bell-flower	77
Alkanet	85	Bent-grass	145
Allseed	46	Betony	99
Anemone	2	Bilberry	78
Angelica	54	Bindweed	82
Apple	43	Bindweed, hooded	82
Archangel	98	Birch	118
Arrow-head	134	Bird-cherry	38
Arrow-grass	134	Bird's-foot	35
Ash	80	Bird's-foot Trefoil	34
Asparagus	129	Bird's-nest	125
Avens	39	Bishop's-weed	51
		Bitter-cress	8
Barberry	5	Bitter-sweet	86
Barley	151	Bitter-vetch	38
Bartsia	89	Blackberry	40
Basil	101	Black Bryony	128
Basil Thyme	100	Blackthorn	38
Bastard Balm	101	Bladderwort	102
Bastard Stone-parsley	51	Blinks	46
Bastard Toadflax	114	Bluebell	129
Beak-rush	139	Bluebottle	70

INDEX TO THE ENGLISH NAMES.

Name	Page		Name	Page
Bog Asphodel	133		Catch-fly	18
Bog-rush	139		Cat-mint	100
Borage	86		Cat's-ear	64
Brakes	164		Cat's-tail	135
Bramble	40		Cat's-tail grass	144
Brome-grass	149		Celandine	6
Brooklime	88		Celery	50
Brookweed	105		Chaffweed	105
Broom	30		Chamomile	77
Broom-rape	87		Charlock	15
Buckbean	82		Cherry	38
Buckler-fern	156		Chervil	55
Buck's-horn Plantain	106		Chestnut	122
Buckthorn	29		Chicory	67
Buckwheat	110		Chickweed	21
Bugle	97		Cinquefoil	41
Bugloss	85		Clary	95
Bugloss, viper's	83		Cleavers	60
Bulrush	140		Clover	32
Burdock	67		Club-moss	169
Bur Marigold	70		Club-rush	140
Burnet	41		Cockle	19
Burnet-leaved Rose	42		Cock's-foot grass	148
Burnet Saxifrage	51		Codlins-and-cream	44
Bur-reed	135		Cole-seed	14
Butcher's Broom	129		Coltsfoot	73
Butter-bur	73		Columbine	5
Butterfly Orchis	126		Comfrey	85
Butterwort	102		Common Mallow	23
			Corn Cockle	19
Cabbage	13		Corn-flag	127
Calaminth	100		Corn Marigold	76
Campion	19		Corn-salad	61
Campion, bladder	18		Cotton-grass	140
Canary-grass	144		Cotton-thistle	69
Carex	141		Couch-grass	151
Carline Thistle	69		Cow-parsnep	54
Carrot	56		Cowslip	103

INDEX TO THE ENGLISH NAMES.

	Page
Cow-wheat	90
Crab-tree	43
Crane's-bill	26
Crested Hair-grass	146
Crosswort	59
Crowfoot	2
Cuckoo-pint	135
Cudweed	72
Currant	48
Daffodil	128
Daisy	75
Dandelion	65
Danewort	58
Darnel	152
Dead-nettle	98
Deptford Pink	18
Devil's-bit	62
Dock	112
Dodder	83
Dog-rose	43
Dog's-tail grass	148
Dog-wood	57
Dropwort	39
Duckweed	136
Dutch Myrtle	118
Dwarf Elder	58
Dwarf Mallow	23
Dyer's Rocket	15
Earth-nut	51
Elder	58
Elecampane	75
Elm	117
Enchanter's Nightshade	44
Eryngo	49
Eye-bright	90

	Page
False Brome-grass	151
Fennel	53
Fenugreek	32
Fern, Lady	158
Fern, Male	156
Fescue-grass	148
Feverfew	76
Figwort	91
Filago	72
Filmy-fern	165
Fir	123
Flax	22
Flax-seed	23
Fleabane	75
Fleawort	74
Flix-weed	12
Flote-grass	147
Flower-de-luce	127
Flowering-fern	166
Flowering-rush	133
Fool's Parsley	53
Forget-me-not	84
Foxglove	91
Foxtail-grass	143
Frog-bit	124
Fumitory	7
Furze	29
Gale	118
Garlic	130
Garlic Mustard	12
Gean	38
Gentian	81
Geranium	26
Germander	97
Germander Speedwell	89
Gipsywort	94
Glasswort	107

INDEX TO THE ENGLISH NAMES. 183

	Page
Goat's-beard	63
Golden Rod	74
Golden Saxifrage	49
Goldilocks	3
Good-King-Henry	108
Gooseberry	48
Goose-foot	107
Goose-grass	60
Gorse	29
Gout-weed	51
Grass-wrack	138
Gromwell	83
Ground-ivy	100
Groundsel	74
Guelder-rose	58
Hair-grass	145
Hard-fern	164
Harebell	77
Hare's-ear	52
Hare's-foot Trefoil	32
Hart's-tongue	162
Hawk-bit	64
Hawk's-beard	65
Hawthorn	43
Hazel-nut	123
Heart's-ease	17
Heath	79
Heath-grass	148
Hedge Mustard	11
Hedge Parsley	56
Hellebore	4
Helleborine	124
Hemlock	54
Hemp Agrimony	72
Hemp-nettle	98
Henbane	86
Herb-Robert	26

	Page
Hog-weed	54
Holly	80
Honewort	50
Honeysuckle	58
Hop	117
Horehound	97
Horehound, white	100
Hornbeam	123
Horned Pondweed	138
Horned Poppy	6
Hornwort	116
Horse-radish	9
Horseshoe Vetch	35
Horse-tail	168
Hound's-tongue	86
Houseleek	47
Iris	127
Ivy	57
Jack-by-the-hedge	12
Kale	15
Knapweed	70
Knawel	110
Kidney-vetch	30
Knot-grass	46, 110
Kobresia	140
Kœleria	146
Ladies'-fingers	30
Lady-fern	158
Lady's-mantle	41
Lady's-smock	8
Lady's-tresses	125
Lamb's Lettuce	61
Lime-tree	24
Linden	24

INDEX TO THE ENGLISH NAMES.

	Page
Ling	79
Live-long	47
Loosestrife	104
Loosestrife, purple	45
Lousewort	91
Lychnis	19
Lyme-grass	151
Madder	59
Maiden-hair	165
Maiden-hair Spleenwort	161
Male-fern	156
Mallow	23
Maple	26
Mare's-tail	45
Marjoram	96
Marsh Marigold	4
Mat-grass	143
Mat-weed	145
May	43
Meadow-grass	147
Meadow Rue	1
Meadow Saffron	130
Meadow-sweet	39
Medick	31
Melic-grass	146
Melilot	31
Mercury	114
Mignonette	15
Milfoil	77
Milk-thistle	68
Milkwort	17
Milkwort, sea	103
Millet-grass	144
Mint	95
Mistletoe	57
Mithridate Mustard	10

	Page
Moonwort	167
Moor-grass	170
Motherwort	97
Moschatel	56
Mountain Ash	43
Mouse-ear Chickweed	21, 22
Mud-rush	139
Mugwort	71
Mullein	93
Musk Mallow	23
Musk Thistle	68
Mustard	14
Narcissus	128
Navelwort or Pennywort	47
Navew	13
Needle Greenweed, or Pettywhin	30
Nettle	116
Nightshade	86
Nightshade, enchanter's	44
Nipple-wort	66
Nit-grass	144
Oak	122
Oat	150
Oat-like grass	146
Onion	130
Orache	108
Orchis	125
Orpine	47
Osier	120
Osmund Royal	166
Ox-eye	76
Ox-lip	103
Ox-tongue	63

INDEX TO THE ENGLISH NAMES.

	Page
Pansey	17
Parsley	50
Parsley-piert	41
Pear	43
Pearlwort	19
Pellitory-of-the-wall	116
Penny-cress	10
Penny-royal	96
Pepperwort	13
Periwinkle	81
Persicaria	110
Pig-nut, or Earth-nut	51
Pilewort	2
Pimpernel	104
Pink	18
Plantain	106
Ploughman's Spikenard	75
Plum	38
Polypody	153
Pond-weed	137
Poor-Man's Weather-glass	104
Poplar	121
Poppy	5, 6
Primrose	103
Privet	80
Purslane, water	45
Quaking-grass	148
Radish	15
Ragged-Robin	19
Ragwort	74
Ramsons	130
Rape	14
Raspberry	40
Red Dead-nettle	98

	Page
Reed	150
Reed-mace	135
Rest-harrow	30
Ribwort	106
Rock-cress	8
Rock-rose	15
Rock-rose, white	16
Rocket	14
Rose	42
Rowan-tree	43
Ruppia	138
Rush	131, 132
Rye-grass	152
Sage	95
Sainfoin	85
Sallow	120
Salad-Burnet	41
Saltwort	110
Saltwort, black	103
Samphire	54
Sandwort	20
Sanicle	49
Sauce-alone	12
Sawwort	67
Saxifrage	48
Scabious	62
Scale-fern	163
Scorpion-grass	84
Scurvy-grass	10
Sea Blite	109
Sea Chamomile	76
Sea Holly	49
Sea Kale	15
Sea Lavender	105
Sea Radish	15
Sea-reed	145
Sea Rocket	11

INDEX TO THE ENGLISH NAMES.

	Page
Sea Spleenwort	161
Self-heal	101
Service-tree	43
Shepherd's-needle	55
Shepherd's-purse	12
Sherardia, or Field Madder	60
Shield-fern	154
Shore-weed	106
Sibthorpia, or Cornish Moneywort	93
Silver-weed	41
Skull-cap	101
Sloe	38
Small-reed	145
Snapdragon	92
Sneezewort	77
Snowdrop	128
Soapwort	18
Soft-grass	146
Sorrel	112
Sorrel-wood	28
Sow-thistle	64
Spearwort	2
Speedwell	88
Spike-rush	139
Spindle-tree	28
Spleenwort	159
Spurge	114
Spurge Laurel	113
Spurrey	43
Spur Valerian	61
Squill	130
Squinancy-wort	60
St. John's-wort	24, 25
Star-thistle	70
Starwort	22
Stitchwort	20, 21

	Page
Stonecrop	47, 48
Stork's-bill	27
Strapwort	170
Strawberry	40
Strawberry-headed Cinquefoil	41
Succory	67
Sundew	17
Sweetbriar	42
Sweet Flag, or Sedge	136
Sweet Gale	118
Swine's-cress	13
Sycamore	26
Tamarisk	45
Tansy	71
Tare	36, 37
Teasel	62
Teesdalia	11
Thale-cress	12
Thistle	68
Thrift	105
Thyme	96
Timothy-grass	144
Toadflax	92
Toothwort	87
Traveller's-joy	1
Treacle Mustard	12
Tree Mallow	23
Trefoil	32, 33
Trichonema	127
Turnip	13
Tutsan	24
Twayblade	124
Valerian	61
Vernal-grass	143
Vervain	102

INDEX TO THE ENGLISH NAMES.

	Page		Page
Vetch	36	White-rot	49
Vetchling	37	White-thorn	43
Violet	16	White Waterlily	5
Viper's Bugloss	83	Whitlow-grass	10
		Whortle-berry	78
Wake-robin	135	Wild Basil	100
Wall-flower	7	Wild Chamomile	76
Wall Pellitory	116	Willow	119
Wall Pepper	48	Willow-herb	44
Wall Rue	162	Winter-cress	8
Wart-cress	13	Woodbine	58
Watercress	9	Woodruff	60
Water Dropwort	52	Wood-rush	132
Water Hemlock	52	Wood Sorrel	28
Water Milfoil	45	Wormwood	71
Water Plantain	134	Woundwort	99
Water Parsnep	53		
Water Purslane	45	Yarrow	77
Water Radish	9	Yellow Pimpernel	104
Water Starwort	115	Yellow-rattle	90
Weasel-snout	98	Yellow Waterlily	5
Wheat, or Wheat-grass	151	Yellow-weed	15
Whin	29	Yellowwort	82
White Dead-nettle	98	Yew	123
White Horehound	100		

FINIS.

PRINTED BY
JOHN EDWARD TAYLOR, LITTLE QUEEN STREET,
LINCOLN'S INN FIELDS.

www.ingramcontent.com/pod-product-compliance
Lightning Source LLC
Chambersburg PA
CBHW032129160426
43197CB00008B/568